THE CHEMICAL SOCIETY
MONOGRAPHS FOR TEACHERS No. 24

Principles of Free Radical Chemistry

J. I. G. CADOGAN, PhD, DSc, FRIC, FRSE

Forbes Professor of Organic Chemistry in the University of Edinburgh

LONDON: THE CHEMICAL SOCIETY

Monographs for Teachers

This is another publication in the series of Monographs for Teachers which was launched in 1959 by the Royal Institute of Chemistry. The initial aim of the series was to present concise and authoritative accounts of selected well-defined topics in chemistry for those who teach the subject at GCE Advanced level and above. This scope has now been widened to cover accounts of newer areas of chemistry or of interdisciplinary fields that make use of chemistry. Though intended primarily for teachers of chemistry, the monographs have proved of great value to a much wider readership, including students in further and higher education.

First published July 1973

Published by The Chemical Society, Burlington House, London W1V 0BN, and distributed by The Chemical Society Publications Sales Office, Blackhorse Road, Letchworth, Herts. SG6 1HN.

Printed by Heffers Printers Ltd Cambridge England

CONTENTS

1. Introduction

Definitions

A free radical is an atom, molecule or complex which contains one or more unpaired electrons. This definition is very broad since it includes species such as nitric oxide NO, the chlorine atom Cl·, free methyl $\cdot CH_3$, and atoms of transition metals which contain an unpaired electron in an inner shell. The last named group is customarily excluded from this definition of a free radical. Free radicals can exist in the gaseous and liquid phases and the principles governing their behaviour apply in both phases. Any differences which appear to occur are usually ones of degree which result from the greater dilution of the radicals in the gas phase. In this book the more complicated case of organic radicals in solution will be considered.

Historical

The term radical originally denoted a group of atoms which remained intact throughout a series of reactions, attempts to isolate these were made in the early nineteenth century; Guy-Lussac, in 1815, for example, thought he had isolated CN. However, two new concepts in the 1860s stopped all such talk about radicals. Thus the vapour density method for determination of molecular weights led to a reassessment (correct, this time) of structures of compounds previously thought to be radicals, *e.g.* $(CN)_2$, while valence theory and the tetravalency of carbon, in particular, seemed to spell the end of ideas about radicals with their abnormal valency requirements. No further attempts to isolate radicals were made during the remainder of the century because they were firmly believed not to exist. Therefore, Gomberg's report, in 1900, of the production of the free triphenylmethyl radical, said to be in equilibrium with hexaphenylethane,

$$Ph_3CCPh_3 \rightleftharpoons 2Ph_3C\cdot$$

was badly received. Disbelief gradually turned into acceptance, culminating in Wieland's definition of radicals as 'complexes of abnormal valency which possess additive properties but do not carry an electrical charge and are not free ions'. Up to this time radicals were only exemplified by the long lived free triphenylmethyl, and

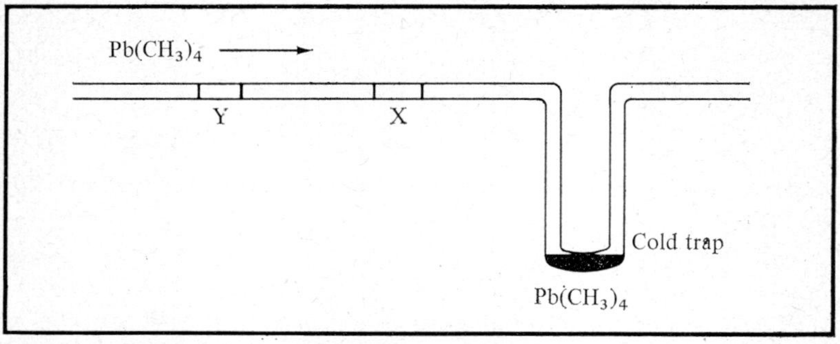

FIG. 1. Demonstration of the existence of methyl radicals in gas phase.

were considered to be rather freakish things having little to do with chemical reactions in general.

Then came Paneth's famous experiment in 1929 in which he and Hofeditz demonstrated the existence of methyl radicals in the gas phase. Gaseous lead tetramethyl was passed through a tube heated at point X (*Fig. 1*). This led to the formation of a lead mirror at point X. Subsequently the tube was heated at point Y, again leading to a deposition of lead at this point. The key observation, however, was that simultaneously with the formation of a mirror at Y, the first formed mirror at X was removed and lead tetramethyl was collected in the cold trap. Paneth and Hofeditz interpreted this to mean that highly reactive methyl radicals capable of attacking lead were being formed on pyrolysis of lead tetramethyl.

That reactive short lived radicals such as aryl could be formed in *solution* was first suggested by D. H. Hey, in 1934, and later confirmed by W. A. Waters. In 1937 Hey and Waters produced a now famous review in which they explained some dozens of hitherto unexplained but well known organic reactions in terms of reactive free radical intermediates. The publication of this review was a turning point in free radical chemistry because before this the concept of the existence of short lived radicals in solutions was little known, and much less appreciated, outside the laboratories of Hey and Waters. Afterwards, as a result of their work and that of Kharasch in the US, the concept gained ground and opened up an entirely new field, of organic chemistry in solution, of great practical and theoretical significance. It is now recognized that an enormous number of oxidation, fragmentation, and ageing reactions, important in technology and biology, proceed *via* radical intermediates; the plastics industry is heavily dependent on the use of free radical polymerization reactions, the US sector alone uses some 30m pounds (at 40c to $1 per pound) of radical initiators *per annum*, while a host of frequently used organic reactions are also now known to involve radicals.

Characteristics of radical reactions

In theory radicals may be formed by homolytic scission (equal splitting) of covalent bonds, in contrast to heterolytic fission which gives ions

Heterolysis $\quad A{:}B \longrightarrow A{:}^- + B^+$ or $A^+ + {:}B^-$

Homolysis $\quad A{:}B \longrightarrow A\cdot + \cdot B$

Whereas the courses of heterolytic or ionic reactions are governed by considerations of charge, polarity, solvation by ionizing solvents *etc.*, homolytic reactions give rise to one valence-electron species which are usually uncharged and therefore not subjected to major polar influences either in the solvent or in the substrate. The major difference between radical and ionic reactions lies, therefore, in the one-electron nature of the former. Radicals react *via* electron pairing steps, involving the free electron at the radical site, to give stable, two-electron bonds, the reactions generally being almost entirely independent of polar factors. The dominance of polar factors in ionic reactions leads to greater differences in activation energy of the possible competing reactions whereas the much diminished importance of polarity in radical reactions leads to the lack of large activation energy differences for the possible competing radical processes. In other words, the operation of polar effects in ionic reactions leads to much more selective reactions in the sense that some reactions occur readily whereas others do not occur at all. Most radical reactions are not restrained in this manner and those polar effects which exist are usually second order and are of major importance only in certain halogenation reactions (Chapter 5) and copolymerizations (Chapter 6). Hence, providing the single electron is not delocalized in the radical (Chapter 3), radicals will react merrily with almost all types of molecules available. Indeed, it is not strictly possible to find a solvent for many radical reactions because such radicals react with the 'solvent' as well as any solutes which may be present. It is clear, therefore, that radical reactions, overall, are potentially much more complicated than ionic processes.

Stability and reactivity

There is often confusion over the use of the terms *stability* and *reactivity*. Stability means the tendency, either greater or smaller, for a molecule, radical or ion to dissociate into smaller fragments, or to rearrange, *via* bond breaking. Reactivity means the ease with which one species undergoes reaction with another. It is meaningless merely to say that a species, be it a radical, ion or molecule is reactive; we must always specify the partner in the reaction and the reaction conditions, because an ion, say, can be reactive towards one substrate but unreactive towards another even under comparable conditions of reaction. For example, in the nitration of aromatic compounds by sulphuric acid–nitric acid mixture the nitrating species is the nitronium ion NO_2^+. This is highly reactive towards benzene at 0°C but is relatively unreactive towards nitrobenzene because it reacts over 100 000 times less readily with nitrobenzene under the same conditions.

The key factors which determine the ease with which a radical reaction occurs are: (*a*) the degree of localization of the single electron at the radical centre; depending on the structure of the radical this may be as high as 100 per cent, such as in methyl $\cdot CH_3$, or much lower if opportunities exist for delocalization over the rest of the radical, such as in benzyl $PhCH_2\cdot$ or triphenylmethyl $Ph_3C\cdot$ (*see* Chapter 3). Clearly, the more localized the free electron, the more pronounced will be the radical behaviour of the species, and in the case of the highly localized methyl radical remarkably few compounds are unreactive towards it. (*b*) The strength of the covalent bonds being broken and formed during the reaction, *i.e.* the energetics, and clearly this embraces the factors outlined in (*a*) above.

The correct uses of the terms stability and reactivity are illustrated best by consideration of four radicals: t-butoxyl ($Me_3CO\cdot$), methyl ($\cdot CH_3$), the substituted nitroxyl radical (**1**), and benzyl ($PhCH_2\cdot$).

Thus, it is correct to say that the t-butoxyl radical is unstable at 120°C in toluene because at 120°C it decomposes to give acetone and the methyl radical $Me_3CO\cdot \rightarrow Me_2CO + \cdot Me$. At 90°C, however, the t-butoxyl radical is stable (because it does not decompose significantly to give acetone and methyl at this temperature) but it is highly reactive towards toluene $PhMe + Me_3CO\cdot \rightarrow PhCH_2\cdot + Me_3COH$. It therefore has a short lifetime despite the fact that it is stable in the sense that, at 90°C, it has little tendency to decompose. It is clearly incorrect, therefore, to merely say that the t-butoxyl radical is stable or, for that matter, unstable, without specifying the conditions.

Similarly it is correct to say that a methyl radical at 50–120°C is stable (because it shows no tendency to decompose; a single methyl radical in the vacuum of outer space would exist indefinitely). The methyl radical nevertheless has very high reactivity, as a result of its highly localized single electron, towards chloroform (or carbon tetrachloride, or toluene, or diethyl ether *etc.*). Ideally, we want to quantify this reactivity in terms of rate constants but such quantitative data are not always available. In the reaction of methyl radicals with chloroform, mentioned above, the half life of the radical is in the region of 10^{-9} s, depending on concentration.

It is also correct to say that the nitroxyl radical (**1**) is stable. Further, it is unreactive towards chloroform (or carbon tetrachloride, or benzene *etc.*). It is also unreactive towards air. In fact it is so unreactive that it is stable indefinitely in the solid state in a reagent bottle on the laboratory shelf. It is reactive only in electron transfer reactions and towards another free radical such as methyl, with

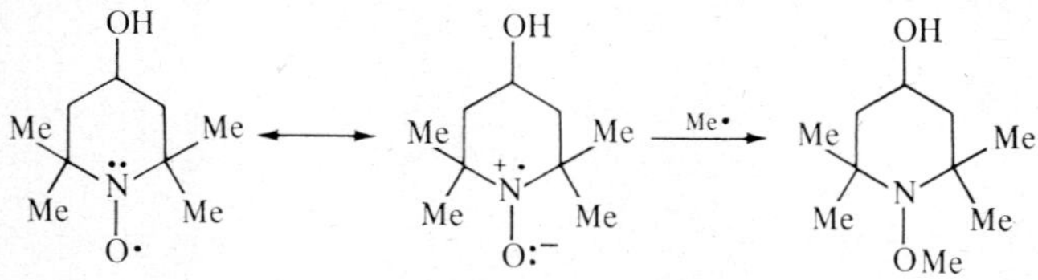

(1) **(2)** **(3)**

which it reacts irreversibly by electron pairing to give a stable methyl hydroxylamine (**3**). In the case of this nitroxyl radical its stability derives from the strong bonds of which it is composed and hence its low tendency to dissociate (or decompose), whereas its unreactivity towards most other species derives from the delocalization of of the odd electron across the N–O bond, as depicted in resonance terms by (**1**) ⟷ (**2**). The fact that such delocalization is sometimes referred to as resonance stabilization has led to the nitroxyl radical being termed 'stable' by virtue of its unreactivity towards other species, rather than by virtue of its lack of tendency to decompose. But, in general, it is far more satisfactory to retain the word stability for use in connection with ease of decomposition or dissociation and to refer to radicals, such as nitroxyls, with a low reactivity towards other species, as being radicals of long life, or long lived.

Finally, it is correct to say that the benzyl radical ($PhCH_2\cdot$) is stable at moderate temperatures (it does not decompose). It is also unreactive towards carbon tetrachloride at 80°C because the single electron is delocalized over the benzene ring with which it is conjugated. The resultant decrease in reactivity is such that in carbon

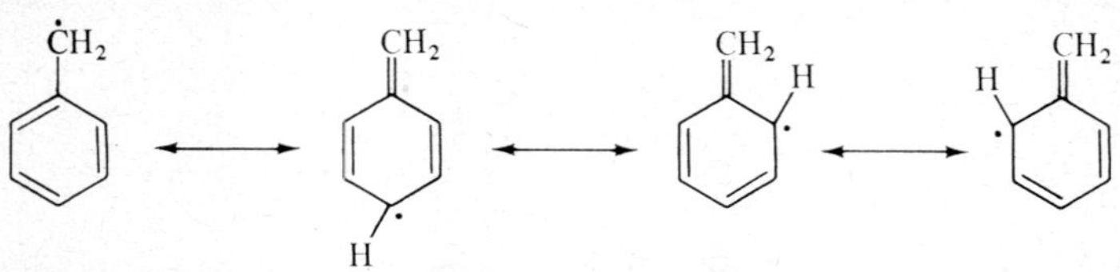

tetrachloride this factor combined with the high C–Cl bond strength*

* A table of bond strengths is given in the Appendix.

is sufficient to prevent reaction occurring. Thus in this solvent the benzyl radical is sufficiently long lived to have a chance of meeting another benzyl radical with subsequent rapid electron pairing to give bibenzyl.

$$PhCH_2\cdot + \cdot CH_2Ph \longrightarrow PhCH_2CH_2Ph$$

On the other hand the Br–CCl_3 bond in bromotrichloromethane is sufficiently weak to react even with the delocalized benzyl radical.

$$PhCH_2\cdot + BrCCl_3 \longrightarrow PhCH_2Br + \cdot CCl_3$$

Obviously it would be incorrect to say that the benzyl radical is unreactive or reactive without specifying the reactants and the conditions.

Strictly, therefore, the reactants and reaction conditions should be specified when referring to the reactivity of radicals. Nevertheless, it is a convenient fact that certain radicals, *e.g.* methyl, phenyl, or alkoxyl react with most organic molecules very readily. These are known as short lived radicals. Others, such as nitroxyls, certain hindered phenoxyls and related radicals, do not react with organic materials under normal conditions and are known as long lived radicals. A third, very broad, category, including the benzyl radical for example, comes somewhere in between, reacting with some but not all organic molecules. Such species are loosely known as radicals of intermediate lifetime.

2. Classification of Reactions of Radicals in Solution

This Chapter provides a brief classification of the types of reactions radicals can undergo. The reader completely unfamiliar with radical reactions may find it easier to omit some of the examples on the first reading and to refer back to these having read the later chapters.

Radical reactions can involve the rupture of a covalent bond in another molecule with formation of a new radical, *e.g.*

$$R\cdot + XY \longrightarrow RX + Y\cdot$$

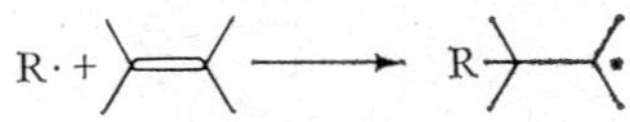

as well as the direct combination of two radical centres, *e.g.*

$$R'\cdot + R\cdot \longrightarrow RR'$$

In the former the new radical is available in turn for further radical reactions of one sort or another. Since reactions of radicals involve single electrons, rather than pairs of electrons as in ionic processes, a need arose for some sort of arrow convention corresponding to the shorthand notation used so successfully in ionic reactions. In the latter case the tail of the arrow ↷ is taken to indicate the point of origin of an electron pair and the head the point of arrival, assuming that the reaction goes to completion. In the case of radical reactions the half arrow notation is often used to indicate the pair or unpairing of *single* electrons (⌒), *e.g.*

This, again, is a useful pictorial aid, but it should be remembered that arrows, whether 'full' or 'half', are nothing more than this.

The following classification divides reactions of radicals into three main types: unimolecular reactions of the radical not involving a further species and bimolecular reactions which involve reaction with another molecule, or radical.

It is important to realize that any combination of the following types of reaction may take place simultaneously and/or consecutively, depending on the nature of the radical, *i.e.* whether localized or delocalized, and the available reactants. Fortunately, by varying the relative concentrations of the reactants it is often possible to manipulate the sequence of events to produce the desired reaction.

I Reactions not involving another species: unimolecular radical reactions

All reactions in this class result from the instability of the first formed radical. In practice the radicals may completely decompose or rearrange before reaction with other molecules or radicals present.

A. Decomposition

Reactions in which the radical decomposes to give a stable molecule and a new radical.

Example IA(*i*). Decomposition of t-butoxyl, formed by thermolysis of di-t-butyl peroxide at 130°C, into acetone and methyl.

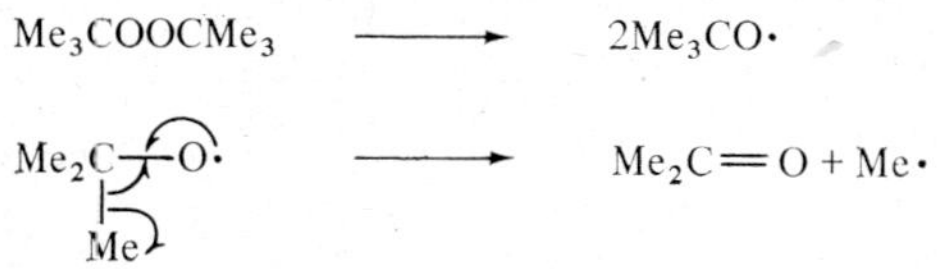

Example IA(*ii*). Decomposition of benzoyloxyl, formed by thermolysis of dibenzoyl peroxide at 80°C, into a phenyl radical and carbon dioxide.

$$\mathrm{Ph\overset{O}{\overset{\|}{C}}OO\overset{O}{\overset{\|}{C}}Ph \longrightarrow 2PhCO_2\cdot}$$
$$\mathrm{PhCO_2\cdot \longrightarrow Ph\cdot + CO_2}$$

Example IA(*iii*). Decomposition of the acetyl radical, formed by reaction of a phenyl radical with acetaldehyde at 80°C, to give methyl and carbon monoxide.

$$\mathrm{Ph\cdot + MeCHO \longrightarrow PhH + Me\dot{C}O}$$
$$\mathrm{Me\dot{C}O \longrightarrow Me\cdot + CO}$$

Example IA(*iv*). Decomposition of the *OOOS*-tetramethyl thiophosphoranyl radical, formed by the sequence shown, into trimethyl phosphorothionate and the methyl radical.

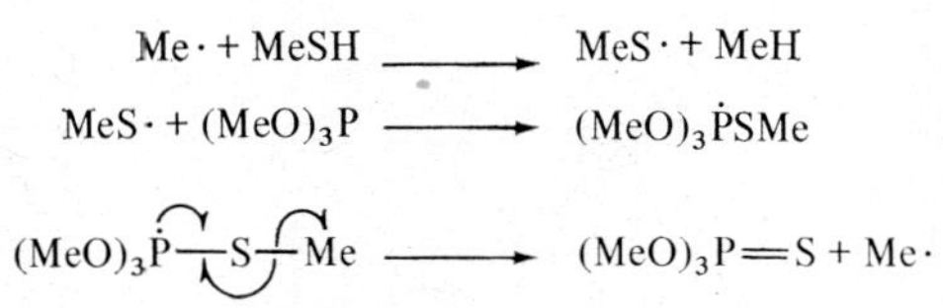

B. Rearrangement

(*a*) Breaking of an adjacent C–C bond in a cyclic system with concomitant formation of a new strong bond, usually carbonyl, and a new, isomeric radical. This type of reaction is the cyclic analogue of *IA*(*i*). (*b*) Migration of an atom, usually a 1,2 or 1,5 shift of hydrogen or halogen, *via* intramolecular abstraction by the radical centre, thus creating a new, isomeric radical.

Example IB(*i*). Rearrangement of the tertiary alkoxyl radical produced by thermolysis of bis(1-methylcyclohexyl)peroxide, into the 6-oxoheptyl radical.

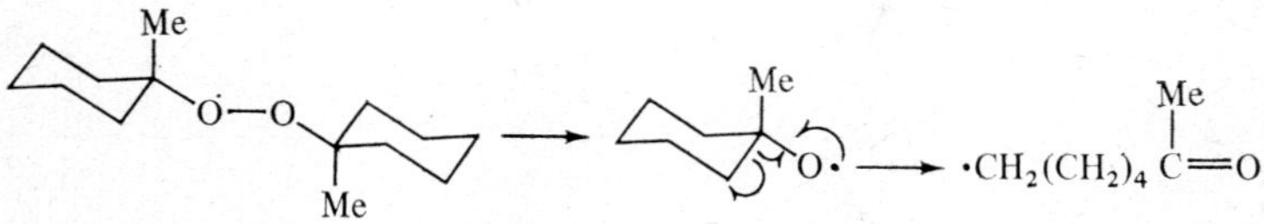

Example IB(*ii*). Rearrangement of the α-tetrahydrofuranyl radical, produced by reaction of a methyl radical with tetrahydrofuran, into the 4-oxobutyl radical.

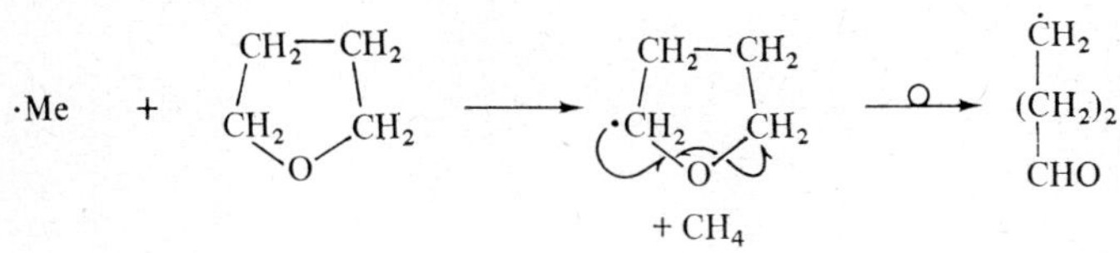

Example IB(*iii*). Rearrangement of the benzylidene acetal radical, produced by reaction of a methyl radical with the parent cyclic acetal, to give an isomeric ester-radical

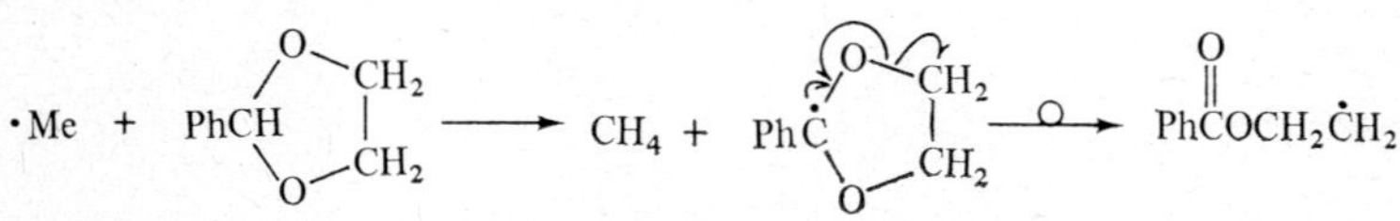

Example IB(*iv*). Rearrangement, *via* 1,2-chlorine shift, of the 1,1,1,4,4,4-hexachloro-2-butyl radical, produced by addition of a trichloromethyl radical to 1,1,1-trichloropropene by the sequence shown:

$$Me\cdot + BrCCl_3 \longrightarrow MeBr + \cdot CCl_3$$

$$\cdot CCl_3 + CH_2{=}CHCCl_3 \longrightarrow Cl_3CCH_2\dot{C}HCCl_3$$

$$\downarrow$$

$$Cl_3CCH_2CHClCCl_2\cdot$$

Example IB(*v*). Rearrangement, *via* 1,5-hydrogen atom shift, of the 6-oxoheptyl radical produced on decomposition of bis(1-methylcyclohexyl peroxide) (*see IB*(*i*) above), to give the isomeric 2-oxo-3-heptyl radical. This reaction, like the succeeding reaction *IB*(*vi*) is facilitated by the ease of formation of a six-membered transition state.

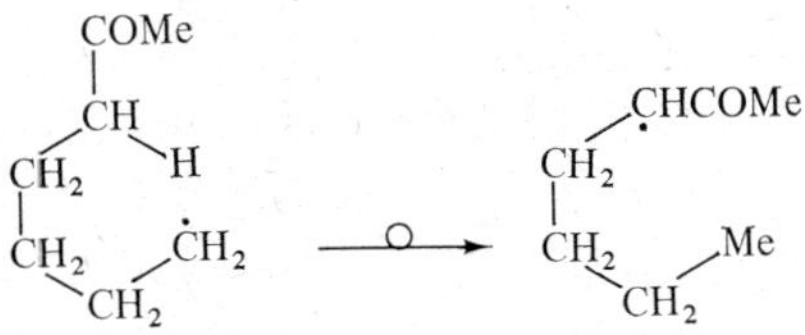

Example IB(*vi*). Rearrangement of the 5-phenylpentyloxyl radical, formed by photolysis of 5-phenylpentyl nitrite, to the isomeric 1-hydroxy-5-phenyl-4-pentyl radical. This is an example of the Barton reaction, discussed later, which is very useful in the synthesis of steroid derivatives.

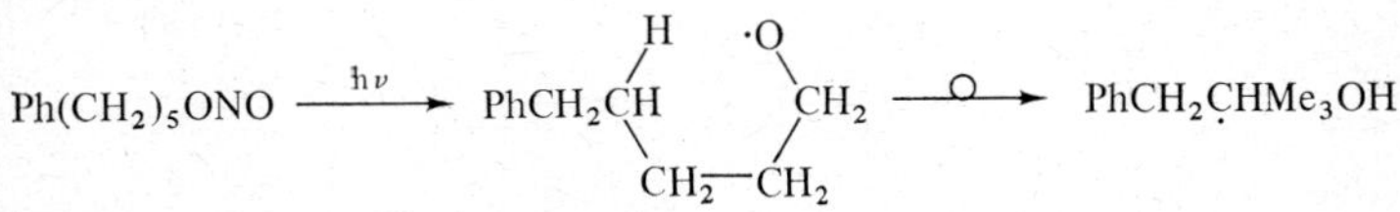

Example IB(*vii*). Rearrangement *via* 1,5-hydrogen shift of the aryl radical shown, formed by decomposition of the corresponding diacyl peroxide.

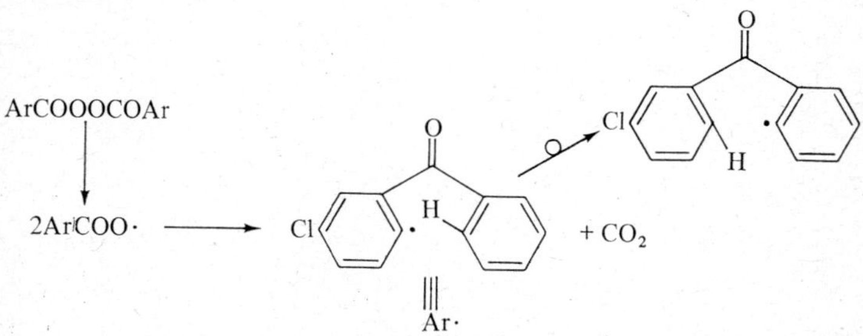

Example IB(*viii*). Rearrangement of the $\beta\beta$-dimethylphenethyl radical.

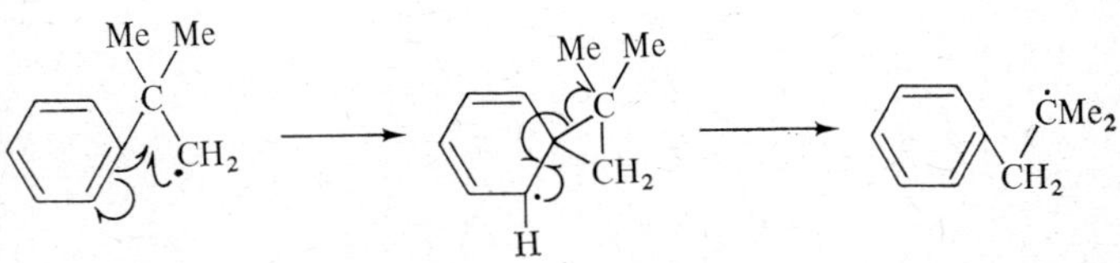

II Radical–molecule interactions

All reactions in this class have a greater probability of occurrence, relative to radical–radical interactions, the lower the concentration of the radical and the higher the concentration of the other reactant, which is often used as 'solvent'. On the other hand, the greater the possibility of delocalization of the odd electron in the radical the smaller the chance of reaction with the solvent, and the greater the chance of radical–radical interactions (*see* below).

A. Addition to unsaturated systems

(*a*) Addition of a radical to an olefinic double bond to give a new, saturated, radical; similar reactions occur with acetylenes. (*b*) Addition of a radical to an aromatic double bond, *e.g.* to benzene, to give a cyclohexadienyl radical.

These reactions are extremely important and are widespread in radical chemistry. Reactions of type (*a*) are the first step in the radical induced polymerization of olefins, while reaction (*b*) is the first step in so-called radical substitutions of aromatic compounds, *i.e.* homolytic aromatic substitution.

Example IIA(*i*). Addition of a trichloromethyl radical, formed in the sequence shown, to oct-1-ene, to give the saturated 1,1,1-trichloro-3-nonyl radical.

$$\text{Me}\cdot + \text{BrCCl}_3 \longrightarrow \text{MeBr} + \cdot\text{CCl}_3$$

$$\cdot\text{CCl}_3 + \text{C}_6\text{H}_{13}\text{CH}{=}\text{CH}_2 \longrightarrow \text{C}_6\text{H}_{13}\dot{\text{C}}\text{HCH}_2\text{CCl}_3$$

Example IIA(*ii*). Addition of a trichloromethyl radical, formed from carbon tetrachloride in the sequence shown, to styrene to give the 1,1,1-trichloro-3-phenyl-3-propyl radical, which, being deactivated by delocalization of the benzylic radical, reacts by further addition reactions with styrene to give a high molecular weight n:1 adduct, rather than by the energetically more expensive pathway of breaking the strong C–Cl bond in carbon tetrachloride.

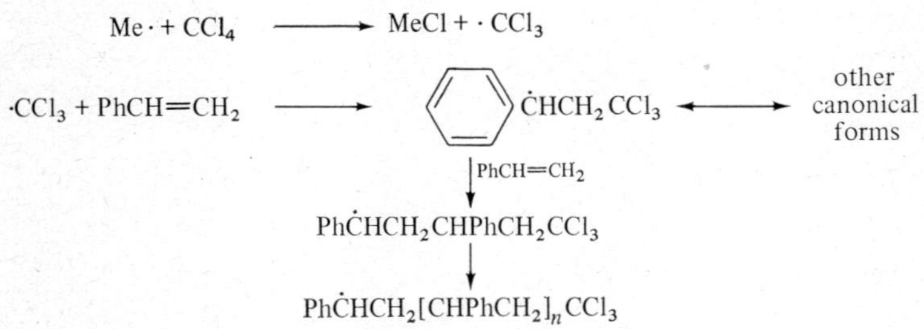

Example IIA(iii). Addition of a *p*-chlorophenyl radical, formed by decomposition of bis(*p*-chlorobenzoyl) peroxide, to benzene to give a substituted cyclohexadienyl radical, the first step in homolytic aromatic substitution.

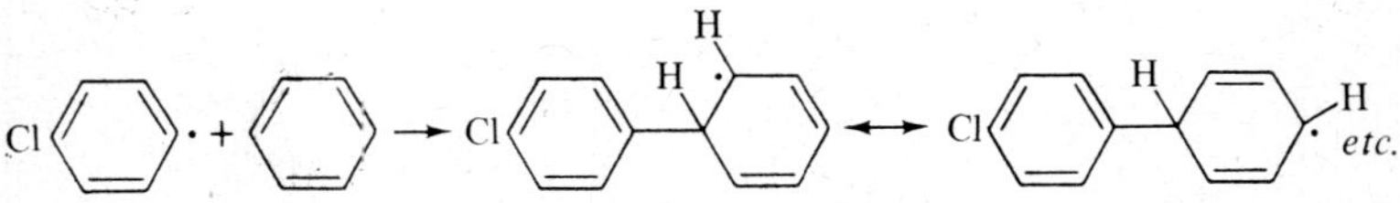

Example IIA(iv). Reaction of vinyl chloride with a 2-cyanopropyl radical, formed by thermolysis of azobisisobutyronitrile, to give an intermediate radical which undergoes further addition with more vinyl chloride to give a polymeric radical, thus leading eventually to polyvinyl chloride (pvc).

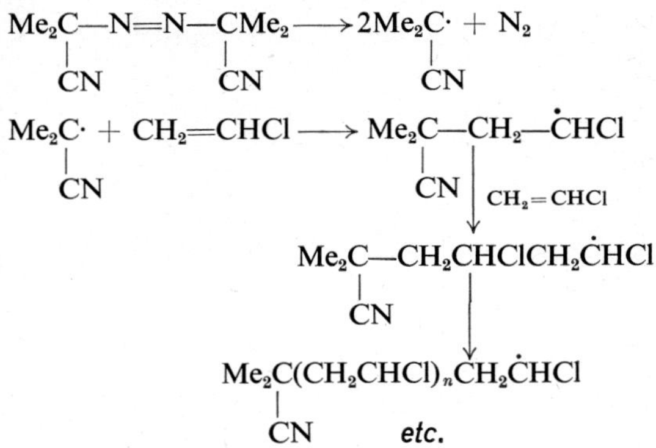

B. Abstraction or displacement: S_H2 reactions

These are bimolecular reactions involving homolytic attack of a radical on a molecule, hence the description S_H2 (substitution homolytic bimolecular). Most of these reactions involve attack by the radical on a *univalent* atom, usually terminal halogen or hydrogen: such processes are often termed abstraction reactions and they give rise to a new radical. (Examples *IIB*(*i*), (*ii*) and (*iii*).) Another group of S_H2 reactions which is becoming increasingly important, involves homolytic substitution at *multivalent* atoms, *e.g.* phosphorus and boron. S_H2 reactions do not normally occur at saturated carbon centres.

S_H2 reactions can be *synchronous via* a *transition state* [R· · ·X· · ·Y].

$$R\cdot + X{-}Y \longrightarrow [R\cdots X\cdots Y]\cdot \longrightarrow R{-}X + Y\cdot$$

or *stepwise via* an *intermediate* [RXY]· of finite lifetime.

$$R\cdot + X{-}Y \longrightarrow [RXY]\cdot \longrightarrow RX + Y\cdot$$

If the decomposition of [RXY]· to RX + Y· occurs faster than the time required for a single vibration (*ca* 10^{-13} s) then [RXY]· is a transition state and not an intermediate. Depending on circumstances it is often possible to detect the intermediate by esr spectroscopy (Chapter 10).

Example IIB(i). Reaction of a phenyl radical with carbon tetrachloride, bromotrichloromethane or chloroform to give a trichloromethyl radical and chlorobenzene, bromobenzene or benzene, respectively.

$$\mathrm{Ph\cdot} + \begin{matrix}\mathrm{CCl_4}\\ \text{or}\\ \mathrm{CBrCl_3}\\ \text{or}\\ \mathrm{CHCl_3}\end{matrix} \longrightarrow \mathrm{\cdot CCl_3} + \begin{matrix}\mathrm{PhCl}\\ \text{or}\\ \mathrm{PhBr}\\ \text{or}\\ \mathrm{PhH}\end{matrix}$$

Example IIB(ii). Reaction of a phenyl radical with butan-1-thiol to give a butan-1-thiyl radical and benzene.

$$\mathrm{Ph\cdot} + \mathrm{BuSH} \longrightarrow \mathrm{BuS\cdot} + \mathrm{PhH}$$

Example IIB(iii). Reaction of a t-butoxyl radical with diethyl malonate to give the diethyl malonyl radical and t-butyl alcohol, or with acetaldehyde to give the acetyl radical.

$$\mathrm{Me_3CO\cdot} + \begin{matrix}\mathrm{CH_2(CO_2Et)_2}\\ \text{or}\\ \mathrm{MeCHO}\end{matrix} \longrightarrow \mathrm{Me_3COH} + \begin{matrix}\mathrm{\cdot CH(CO_2Et)_2}\\ \text{or}\\ \mathrm{Me\dot{C}O}\end{matrix}$$

Example IIB(iv). Displacement of an alkyl radical from a trialkylborane by means of a t-butoxyl radical produced by photolysis or thermolysis of di-t-butyl hyponitrite.

$$\mathrm{Bu^tON{=}NOBu^t} \longrightarrow \mathrm{Bu^tO\cdot} + \mathrm{N_2}$$

$$\mathrm{Bu^tO\cdot} + \mathrm{R_3B} \longrightarrow [\mathrm{Bu^tOBR_3}]\cdot \longrightarrow \mathrm{Bu^tOBR_2} + \mathrm{R\cdot}$$

Example IIB(v). Formation of trifluoromethyltrimethyltin by photo-irradiation of a mixture of hexamethylditin and trifluoromethyl iodide.

$$\mathrm{Me_3SnSnMe_3} \longrightarrow \mathrm{2Me_3Sn\cdot}$$

$$\mathrm{Me_3Sn\cdot} + \mathrm{CF_3I} \longrightarrow \mathrm{Me_3SnI} + \mathrm{CF_3\cdot}$$

$$\mathrm{\cdot CF_3} + \mathrm{Me_3SnSnMe_3} \longrightarrow \mathrm{Me_3(CF_3)\dot{S}nSnMe_3}$$

$$\longrightarrow \mathrm{Me_3SnCF_3} + \mathrm{\cdot SnMe_3}$$

Example IIB(vi). Formation of an alkyl radical by photolysis of di-t-butylperoxide in the presence of a trialkylphosphine. In this

$$\mathrm{Bu^tO\cdot} + \mathrm{R_3P} \longrightarrow [\mathrm{Bu^tOPR_3}]\cdot \longrightarrow \mathrm{Bu^tOPR_2} + \mathrm{R\cdot}$$

case the intermediate phosphoranyl radical can be detected by esr spectroscopy (Chapter 10) and it is interesting to compare this case with that of the corresponding thiophosphoranyl radical in example *IA*(*iv*) which *decomposes* by fission of the weaker *S*-alkyl bond.

C. Reaction with oxidizing agents

Radicals readily undergo 1-electron oxidations with oxidizing reagents of suitable redox potential to give positive ions.

Example IIC(*i*). Oxidation of phenyl radicals *via* cuprous ions, as in the Sandmeyer reaction (*see* Chapter 8).

$$Ph\cdot + Cu^{+} \longrightarrow Ph^{+} + Cu$$

Example IIC(*ii*). Oxidation of cinnamyl derived radicals by cupric ions as in the Meerwein reaction (*see* Chapter 8).

$$Ph\dot{C}HCHRCO_2Et + Cu^{2+} \longrightarrow Ph\overset{+}{C}HCHRCO_2Et + Cu^{+}$$

Example IIC(*iii*). Oxidation of hyphosphorous radicals by the diazonium cation as in the Kornblum deamination reaction (*see* Chapter 8).

$$ArN_2^{+} + \cdot P(OH)_2 \longrightarrow Ar\cdot + N_2 + {}^{+}P(OH)_2$$

Example IIC(*iv*). Oxidation of substituted cyclohexadienyl radicals (the intermediate in homolytic aromatic substitution) as described below and in *IIA*(*iii*), by dibenzoyl peroxide.

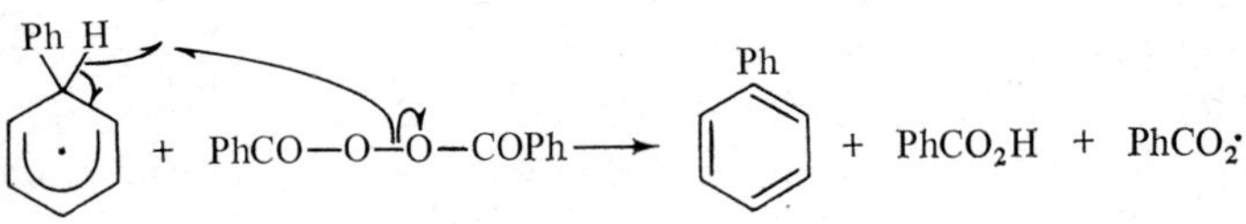

III Radical–radical interactions

The probability of reactions of this type occurring is increased by (*a*) high concentration of the radicals, (*b*) delocalization of the single electron over the radical, thus increasing the lifetime of the radical, (*c*) the absence of weak bonds in any other molecules present with which the radical could interact. Highly localized radicals, such as methyl, or phenyl, react so readily with almost all covalent bonds that the chances of dimerization, in solution, are very small. One of the most widespread misconceptions about radical behaviour is the belief that dimerization is the characteristic reaction of *all* radicals. Only delocalized radicals have a high probability of dimerization in solution.

A. Dimerization or radical coupling

In the general case this becomes

$$R'\cdot + R''\cdot \longrightarrow R'\text{—}R''$$

When $R' = R''$ the reaction is dimerization and when $R' \neq R''$ the reaction is radical coupling or combination. Unless both radicals are extensively delocalized, these are reactions of negligible activation energy.

Example IIIA(*i*). Dimerization of benzyl radicals, formed by reaction of t-butoxyl radicals with toluene at 100°C, to give bibenzyl.

$$Me_3CO\cdot + PhCH_3 \longrightarrow Me_3COH + PhCH_2\cdot$$
$$2PhCH_2\cdot \longrightarrow PhCH_2CH_2Ph$$

In this case toluene is used in excess and the localized, and reactive, t-butoxyl radical reacts very readily by hydrogen abstraction to give t-butyl alcohol and the delocalized benzyl radical, which exists long enough to meet another benzyl radical. The chance of collision of a benzyl radical with a t-butoxyl radical, on the other hand, is very small as a result of great ease of reaction between t-butoxyl and toluene, which is present in excess.

Example IIIA(*ii*). Dimerization of ethoxycarbonylmethyl radicals—formed by reaction of methyl radicals from diacetyl peroxide—with ethyl acetate to give diethyl succinate.

$$MeC(=O)\text{—}O\text{—}O\text{—}C(=O)Me \longrightarrow 2MeC(=O)\text{—}O\cdot \longrightarrow Me\cdot + CO_2$$
$$Me\cdot + CH_3CO_2Et \longrightarrow MeH + \cdot CH_2CO_2Et$$
$$2\cdot CH_2CO_2Et \longrightarrow (CH_2CO_2Et)_2$$

Example IIIA(*iii*). Dimerization of trichloromethyl radicals, formed as in *IIB*(*i*), to give hexachloroethane.

$$2\cdot CCl_3 \longrightarrow Cl_3CCCl_3$$

Example IIIA(*iv*). Dimerization of methyl radicals, formed by the electrolysis of aqueous potassium acetate (the Kolbe process), to give ethane.

$$MeCO_2^-K^+ \longrightarrow MeCO_2\cdot + K$$
$$MeCO_2\cdot \longrightarrow Me\cdot + CO_2$$
$$2Me\cdot \longrightarrow MeMe$$

The reaction is oversimplified here because it involves the electrode, and is thus heterogeneous. This is a factor in the unusually high efficiency of the dimerization of methyl radicals in this case. The high HO–H bond strength of water (Appendix) also makes the alternative hydrogen abstraction reaction less competitive.

Example IIIA(v). Coupling of benzyl and *p*-methylbenzyl radicals, formed by reaction of t-butoxyl radicals with an equimolar mixture of toluene and *p*-xylene, to give 4-methylbibenzyl. Bibenzyl and 4,4′-dimethylbibenzyl are also formed, as in

$$2Me_3CO\cdot + PhCH_3 + p\text{-}C_6H_4(CH_3)_2$$
$$\longrightarrow PhCH_2\cdot + p\text{-}MeC_6H_4CH_2\cdot + 2Me_3COH$$

$$PhCH_2\cdot + p\text{-}MeC_6H_4CH_2\cdot \longrightarrow p\text{-}MeC_6H_4CH_2CH_2Ph$$

Example IIIA(vi). Coupling of trichloromethyl radicals with 1,1,1-trichloro-3-nonyl radicals produced as in *IIA*(*i*) (p 11) to give

$$\cdot CCl_3 + C_6H_{13}\dot{C}HCH_2CCl_3 \longrightarrow C_6H_{13}CH(CCl_3)CH_2CCl$$

1,1,1-trichloro-3-trichloromethylnonane.

B. Radical disproportionation

This involves the collision of two radicals resulting in the abstraction of an atom, usually hydrogen, by one radical from the other, leading to two stable molecules. The atom abstracted is usually β to the radical centre. The reaction derives its driving force from the formation of two new strong bonds and from the fact that β-CH bonds in radicals are usually weak (Appendix).

Example IIIB(i). Disproportionation of two phenylethyl radicals to give styrene and ethylbenzene.

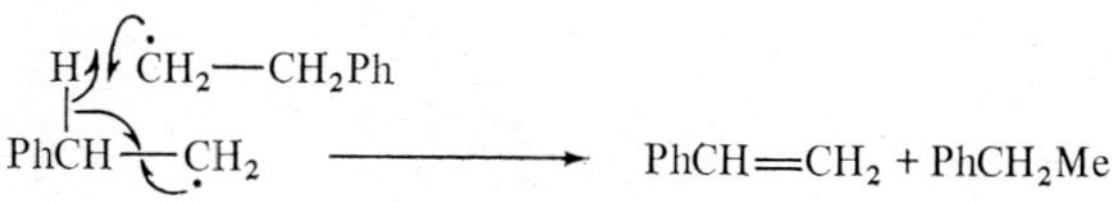

Example IIIB(ii). Disproportionation of two cyclohexyl radicals, formed by reaction of phenyl radicals with cyclohexane, to give cyclohexane and cyclohexene.

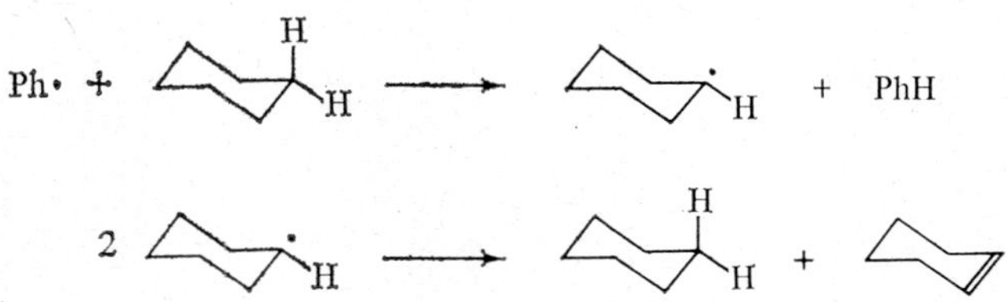

Radical initiation, propagation, termination, and chain reactions

Radical initiation reactions. In the classification above, all decompositions occurring by thermolysis or photolysis of weak bonds, *e.g.* peroxides, azo compounds, nitrites *etc.*, to give radicals are radical initiation reactions, *e.g.* those occurring in examples *IA*(*i*), *IA*(*ii*), *IB*(*i*), *IB*(*vi*), *IB*(*vii*) and *IIIA*(*ii*).

Radical propagation reactions. With two possible exceptions all the reactions involving radical decomposition, radical rearrangement and radical–molecule interactions described above (*i.e.* all examples *IA*, *IB*, *IIA*, *IIB*, *IIC* with the exception of *IIC*(*i*) and *IIC*(*ii*)) involve the formation of a new radical, *i.e.* the reaction is kept 'alive'. The new radical in each case has, in turn, all the possible modes of reaction open to it.

The exceptions, reactions *IIC*(*i*) and *IIC*(*ii*), involve oxidation of the organic radical by a multivalent metal to a carbonium ion, during which cuprous, for example, becomes reduced to copper(0). In certain cases such as the Sandmeyer reaction, however, the copper(0) species is capable of re-entering the reaction by donating an electron to regenerate a cuprous ion. Thus it could be said that a radical centre is being maintained in these cases also, *e.g.* in the formation of phenyl radicals by the reaction of copper(0), in catalytic amount, with the benzene diazonium cation, and subsequent phenylation of benzene.

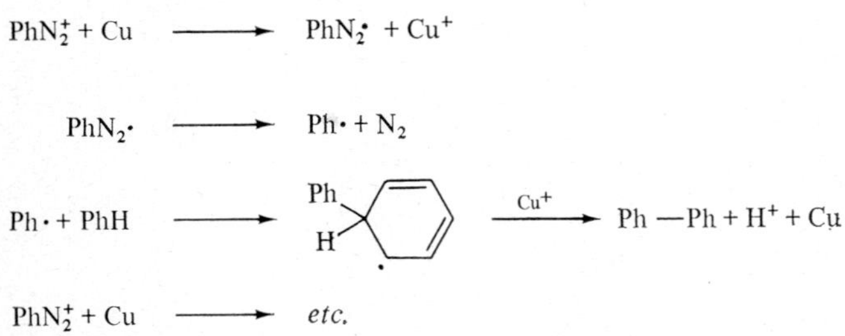

Thus, this is a chain reaction (*see* below) sustained by regenerated copper(0), which is acting as a one-electron donating reagent.

Radical termination. All radical–radical interactions described in the classification are termination reactions in the sense that radical centres are destroyed (*i.e.* all examples *IIIA* and *IIIB*). All radical processes must end by one or more termination reactions of the type outlined, regardless of the number and type of consecutive or concurrent initiation or propagation reactions participating in the overall reaction sequence.

Chain reactions. Many reactions involving radicals are termed chain reactions. Examples of these are halogenation of alkanes (Chapter 5) and radical addition of hydrogen bromide to alkenes (Chapter 6). In these cases the initiating radical, *e.g.* the bromine atom in the addition of HBr to an alkene, is regenerated at a later stage of the reaction, thus leading to a chain sequence,

$$Br\cdot + RCH{=}CH_2 \longrightarrow R\dot{C}HCH_2Br$$
$$R\dot{C}HCH_2Br + HBr \longrightarrow RCH_2CH_2Br + Br\cdot \textit{ et seq.}$$

In theory, therefore, only one initiating bromine atom is required for complete consumption of the reactants, assuming equimolar quantities of these: but in practice it is impossible to introduce only one bromine atom.

3. Radicals of Long and Intermediate Lifetimes

In dealing with radicals of long and intermediate lifetimes it must be emphasized, again, that the use of the term long life is arbitrary. It is usually applied to those radicals whose lifetime in solution at room temperature, in the absence of air or other radicals of 'short' lifetime, is greater than a few minutes. The lifetime of a radical under a given set of conditions depends, in part, on the degree of localization of the electron at the radical centre—sometimes known as the spin density—and such lifetimes can be very long, even years, depending on the structure of the radical and the nature of available reactants. Steric hindrance extends to lifetime.

As is clear from their description, such radicals do not readily enter into reaction with fully bonded molecules, but they can undergo electron pairing reactions, particularly with more highly localized —*i.e.* short lived—radicals. Hence many long lived radicals are useful as radical scavengers, or as inhibitors of radical chain reactions involving the transient existence of localized radicals.

In general any radical of the type (**4**) which has the single electron nominally situated on an atom directly attached to a number of aromatic systems is likely to have a long life. These include triarylmethyl-, diarylamino-, and certain aryloxyl-radicals. Nitroxides, $R_2NO\cdot$, are another particularly important class of long lived radicals.

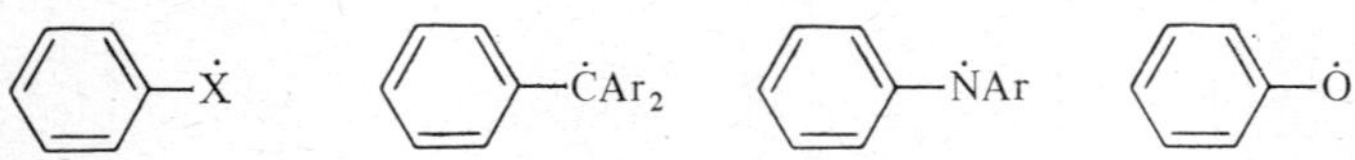

(**4**)

Carbon centred long lived radicals

Triarylmethyl radicals. Gomberg, in 1900, attempted to prepare hexaphenylethane by the reaction of triphenylmethyl chloride with silver.

$$2Ph_3CCl + 2Ag \longrightarrow Ph_3CCPh_3 + 2AgCl$$

Instead of the expected product he obtained a yellow oil of intermediate molecular weight which combined readily with oxygen, iodine, or nitric oxide, to give triphenylmethyl-peroxide, -iodide or nitrosotriphenylmethane respectively. Gomberg explained this in terms of a reversible dissociation of hexaphenylethane into free triphenylmethyl radicals.

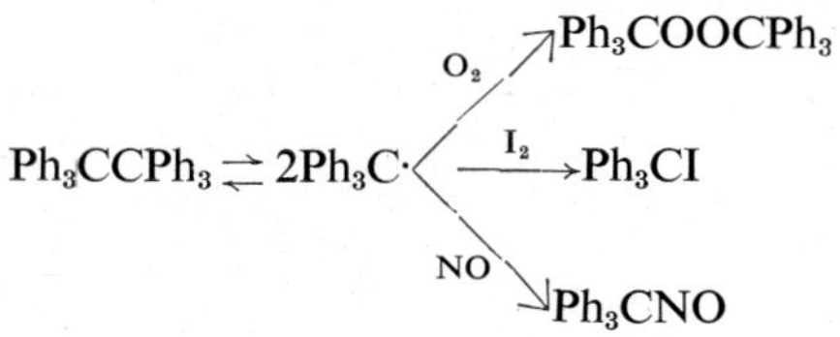

This was greeted with scepticism and, in some quarters, derision, but subsequent esr, spectroscopic, magnetic susceptibility and molecular weight determinations supported the existence of long lived triphenylmethyl radicals.

The main reason for its long life is the possibility of extensive delocalization of the single electron over all three benzene rings. In classical terms the formula of triphenylmethyl is written with a single electron localized on the benzylic carbon atom. It is more profitable to think of the radical in either valence bond (resonance) or molecular orbital terms. The valence bond treatment visualizes a resonance hybrid to which there are 10 acceptable contributing canonical forms, nine of which involve the notional positioning of the odd electron at *o*- or *p*-positions in the peripheral benzene rings rather than at the benzylic carbon atom.

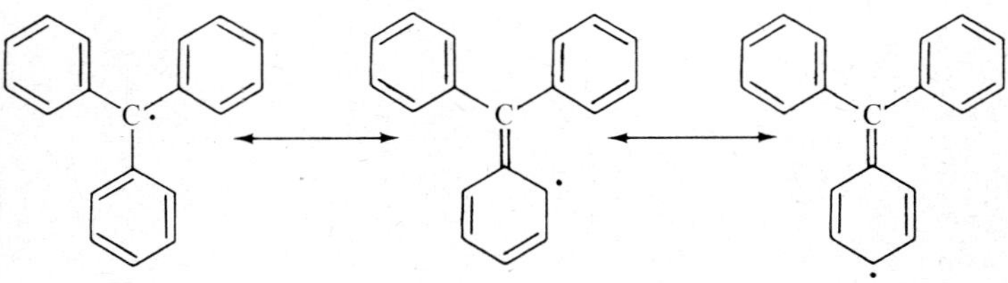

Alternatively, in molecular orbital terms, it can be considered that the benzylic free electron is in a *p*-orbital which overlaps with the π-electron system of all three benzene rings allowing, as in the valence bond representation, extensive delocalization of the single electron: or, put another way, leading to a significant decrease of

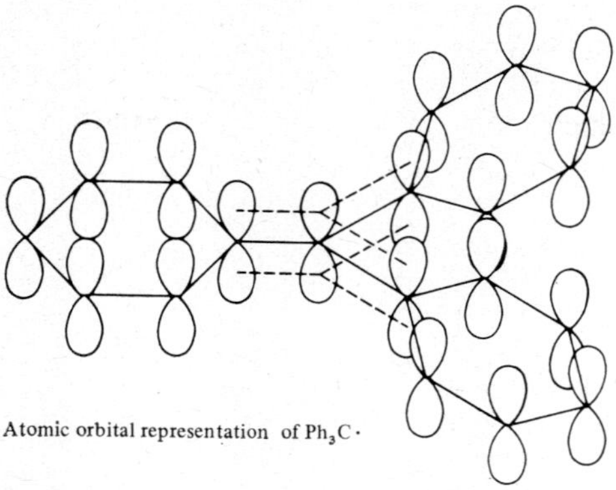

Atomic orbital representation of $Ph_3C\cdot$

spin density at the classical radical centre, and to an increase of spin density at *o*- and *p*-positions of the three benzene rings. This extensive delocalization decreases radical reactivity and increases the thermodynamic stability of the radical compared with its dimer, *i.e.* $R\cdot + R\cdot \rightleftharpoons R–R$. In other words, the C–C bond in the dimer is weak, thus leading to ease of fission into the component radicals, the weakness of the bond resulting from the possibility of delocalization of the electrons in the bond into the remainder of the molecule.

In practice, orbital overlap (or resonance delocalization) is not at a maximum because the radical is forced into a slight propellor conformation by steric hindrance. This is enhanced by large *o*-groups. On the other hand, the presence of *p*-conjugated groups, *e.g.* phenyl, leads to further extensive delocalization and, in this case, to an infinitely longer lifetime.

Compelling experimental evidence for delocalization in the benzene rings of triphenylmethyl was found as recently as 1968. Until then everyone had accepted that dimerization of the triphenylmethyl radical did, in fact, give hexaphenylethane, despite the fact that such a reaction must be considerably sterically hindered.

$$2Ph_3C\cdot \rightleftharpoons Ph_3CCPh_3$$

Nuclear magnetic resonance studies showed conclusively, however, that the dimer is unsymmetrical, formed by the sterically more favourable reaction at a *p*-position of high spin density. So, to date, hexaphenylethane is an unknown compound.

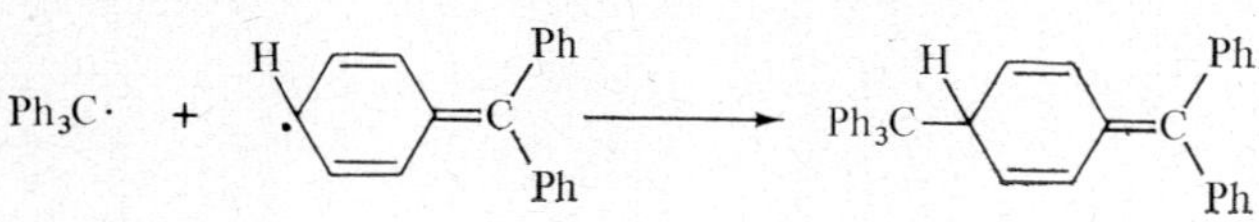

Clearly, the greater the possible extent of delocalization through relatively coplanar conjugated systems the longer the life of the radical. Two good examples of this are the blue perinaphthyl radical (**5**), which exists indefinitely in solution in the absence of

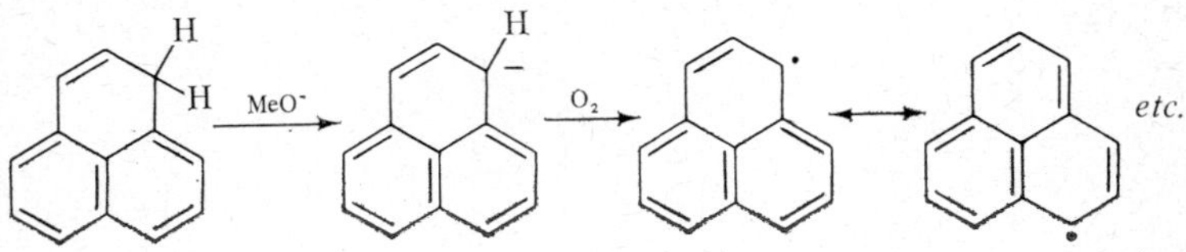

(**5**)

air, and Koelsch's free radical (**6**). The latter red-brown radical is so unreactive that it exists as a crystalline solid unchanged even in oxygen, and such was the disbelief surrounding early work in free radical chemistry that when Koelsch submitted his paper for publication in 1932 it was rejected on the ground that his product could not be a radical. In 1957 Koelsch resubmitted the paper without alteration, pointing out that not only was the original paper correct but that he still had the original sample of the radical unchanged on his laboratory shelf. The paper was accepted.

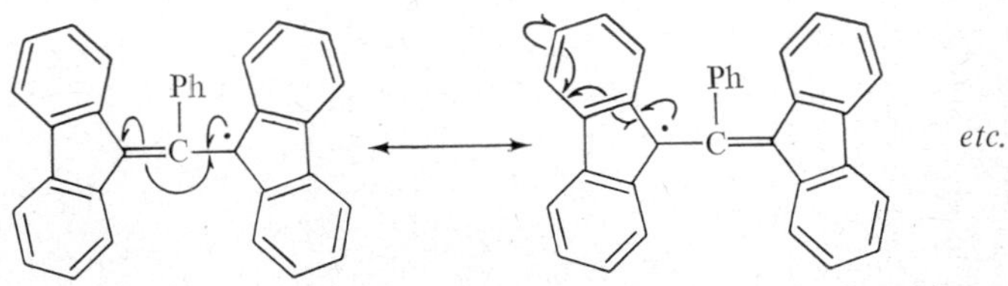

(**6**)

Benzyl and related radicals. In the series $Ph_3C\cdot$, $Ph_2CH\cdot$ and $PhCH_2\cdot$, as the extent of possible delocalization is decreased, *e.g.* in diphenylmethyl or benzyl radicals, so the radicals become more short lived. So much so that although the benzyl radical is sufficiently delocalized to be unreactive towards benzene it does not exist long enough, without dimerization, to give sufficient concentration to be detected easily by esr spectroscopy in a static system.

$$PhCH_2\cdot \xrightarrow{PhH} PhCH_2CH_2Ph$$

Benzyl radicals therefore fall into that rather ill defined group of radicals qualitatively known as radicals of intermediate lifetime, which is the best that can be done in the absence of quantitative data in solution.

Nitrogen centred radicals

Diarylamino radicals are the nitrogen analogues of triarylmethyls and are also long lived because the single electron is delocalized

$$Ph_2NH \xrightarrow{\text{oxidation}} Ph_2NNPh_2 \rightleftharpoons Ph_2N\cdot$$

over two benzene rings. In this case the relatively lower bond strength of the N–N bond helps the dissociation. The phenothiazine radical and the phenothiazine cation radical are closely related, the latter formed by oxidation of phenothiazine in acid solution.

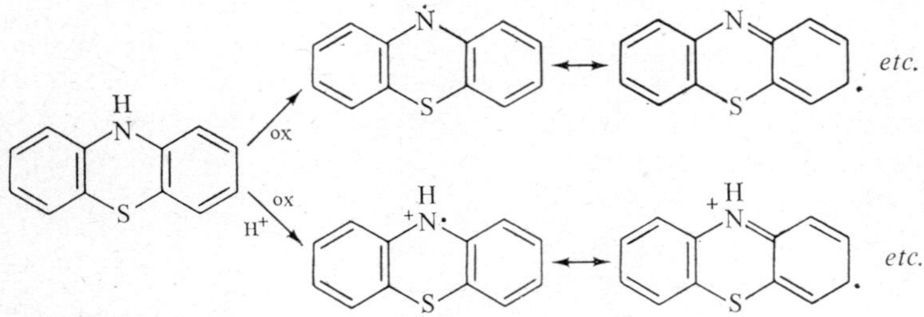

The most important of the long lived nitrogen centred radicals, however, is the highly conjugated, and delocalized, diphenylpicrylhydrazyl. This is a stable, violet black solid, easily obtained by oxidation of the corresponding hydrazine.

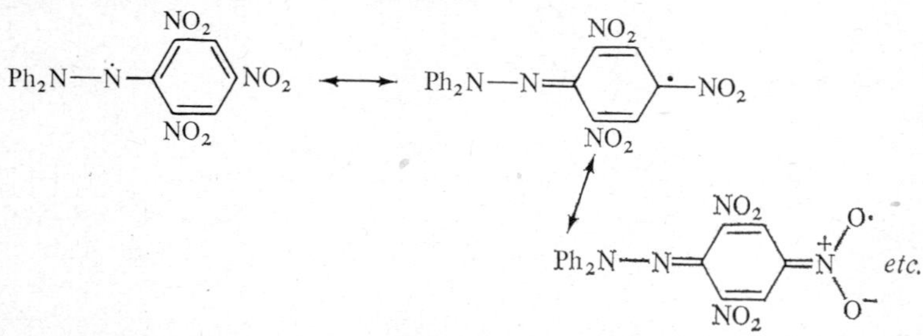

This radical is a very useful radical scavenger, because it is inert towards most molecules but rapidly undergoes pairing reactions with a wide variety of short lived radicals. Since it is itself coloured with a distinctive spectrum, while its combination products are much less so, it can be used to follow rates of radical reactions, spectroscopically, and much quantitative information on radical reactions has been obtained this way, although this radical is now being superseded in this connection by Galvinoxyl (below).

Delocalization of a nitrogen centred radical plays an important part in determining the potency of paraquat (1,1′-dimethyl-4,4′-bipyridylium dibromide) as a weed killer. This agent has almost ideal properties for use in non-tillage systems. It kills all green tissue to which it is applied, it is instantaneously rendered inactive on contact with soil and, although toxic to humans, it is excreted very rapidly by animals with no build-up into meat or excretion into milk.

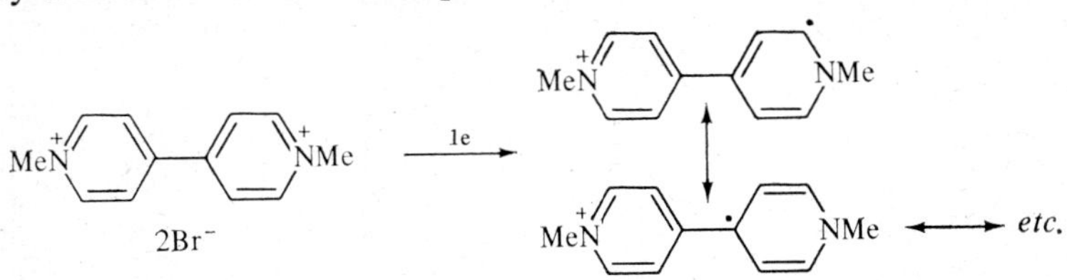

In the plant, paraquat becomes biochemically reduced *via* a one-electron process to give the delocalized nitrogen centred ion radical shown, which exists long enough to react with molecular oxygen, regenerating paraquat and forming a peroxide radical, which is responsible for killing the plant.

Oxygen centred radicals

Aroxyl radicals complete the series $Ar_3C\cdot$; $Ar_2N\cdot$; $ArO\cdot$. They can be prepared very easily by oxidation of the corresponding phenols with manganese dioxide, potassium persulphate or potassium ferricyanide.

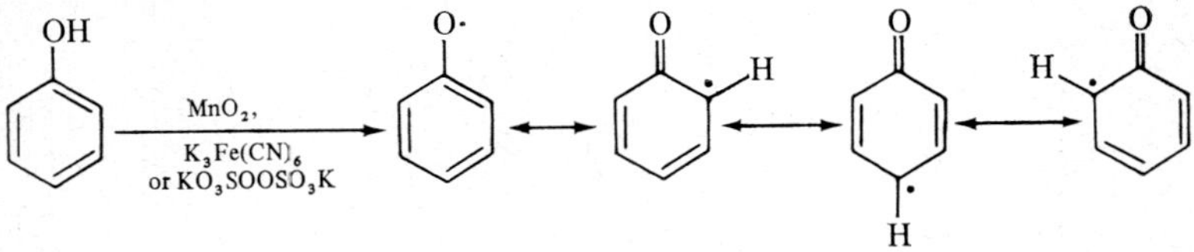

However, the lifetimes of simple aryloxyl radicals such as phenoxyl, *p*-cresoxyl *etc.* are too short for them to fall into our arbitrary class of long lived radicals. Nevertheless, the degree of delocalization is sufficient to reduce their reactivity towards benzene, say, so that they exist long enough to meet and dimerize, *e.g.* as in the first

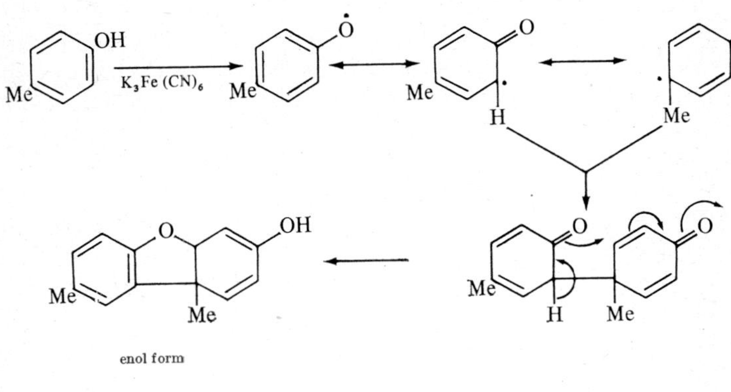

step of the formation of Pummerer's ketone (Chapter 8) on oxidation of *p*-cresol. In this case coupling to give the strong C–C bond is favoured.

The lifetime of aryloxyl radicals can be extended into the long-life ranges by the simple process of blocking the available coupling positions with bulky groups, *e.g.* 2,4,6-tri-t-butylphenoxyl, which is a deep blue solid, m.p. 96–98°C.

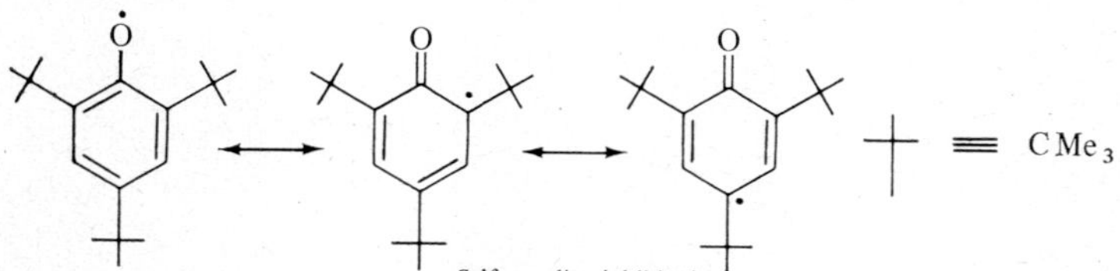

Self coupling inhibited
by steric hindrance

The resulting sterically hindered radical can undergo coupling only with much smaller and probably short lived radicals, and with oxygen.

Hydroquinone is also used as a radical inhibitor. Reaction with a short lived radical leads first to an aryloxyl radical and then to benzoquinone.

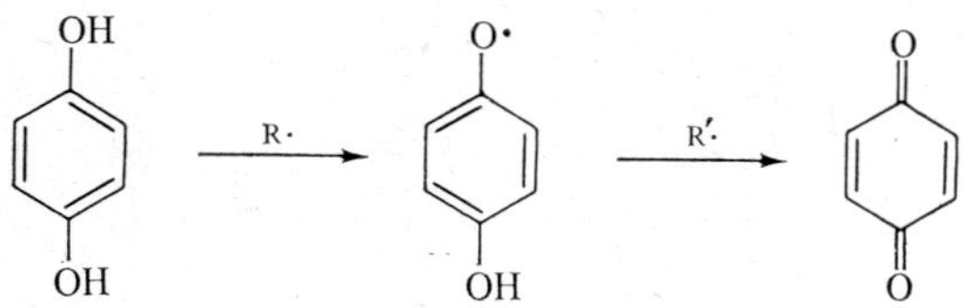

A further very useful member of this class is the long lived aryloxyl (**7**),which is known, fortunately, by the nickname Galvinoxyl, rather than by the systematic 2,6-di-t-butyl-α-(3,5-di-t-butyl-4-oxo-2,5-cyclohexadien-1-ylidene)-*p*-tolyloxyl.

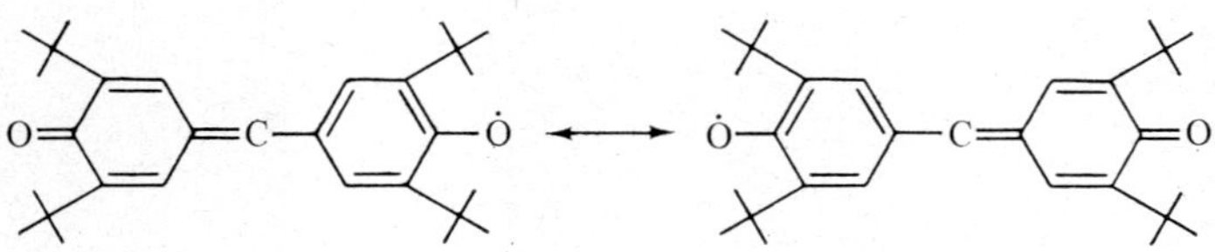

(7)

+ *o*-, *p*-, and methine forms

2,4,6-Tri-t-butylphenoxyl and Galvinoxyl, particularly, are used as radical scavengers and inhibitors of radical chain reactions. Conversely, the fact that a given reaction is inhibited by these species can be taken as a pointer that the initial reaction is homolytic in nature.

Nitroxides

Nitroxides $R_2NO\cdot$, like nitric oxide, are radicals in the sense that they have an unpaired electron. When R is t-alkyl or aryl they are extremely long lived as a result of the very effective delocalization of the electron in the N–O bond, and the lack of α-H atoms.

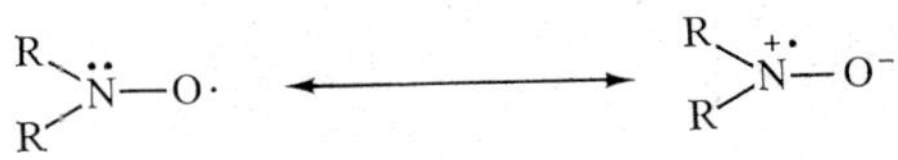

This has also been represented in terms of a σ bond plus a 3-electron bond between the N and O atoms, while Linnett accounts for the properties of the related nitric oxide in terms of his double octet model in which six electrons of one spin are arranged as in (**8**) and the remaining five, of the other spin, as in (**9**). Each atom has a complete octet, therefore, and there are five bonding electrons, with little electronic repulsion. Hence dimerization is energetically unfavoured because it would lead to an increase in interelectronic repulsion without any corresponding increase in the number of bonding electrons.

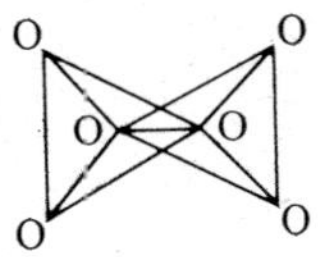

[tetrahedra with a shared edge]

(**8**)

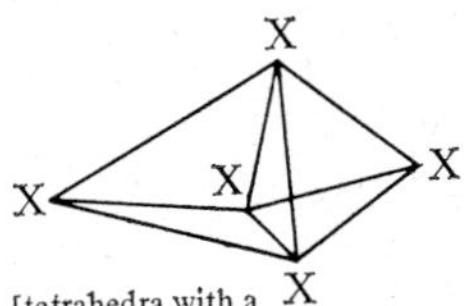

[tetrahedra with a shared face]

(**9**)

Typical examples of nitroxide radicals, which do not dimerize in solution or the solid state, are Frémy's salt (**10**), diphenylnitroxide (**11**) or the cyclic nitroxide (**12**), where steric hindrance at the NO function also contributes to its long life.

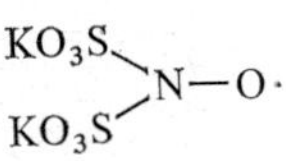

(**10**)

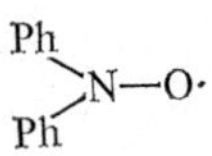

(**11**)

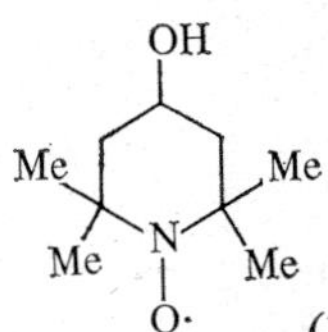

(**12**)

They are easily prepared by oxidation of *NN*-disubstituted hydroxylamines or their salts, by a wide variety of reagents—$KMnO_4$, PbO_2, $K_3Fe(CN)_6$ *etc.*—or by the trapping of short lived radicals by nitroso compounds (*see* Chapter 10: spin trapping).

$$R_2NOH \xrightarrow{\text{ox}} R_2N{-}O\cdot$$

$$RN{=}O + R'\cdot \longrightarrow \begin{matrix} R \\ R' \end{matrix}\!\!>\!N{-}O\cdot$$

Certain nitroxides, such as (**12**), are very useful radical scavengers and inhibitors. Finally, on a point of nomenclature, nitroxides are sometimes known as nitroxyls or nitroxyl radicals.

4. Radical Initiators: Formation of Short Lived Radicals

The simplest way to produce radicals is by direct homolysis of covalent bonds either by thermolysis or photolysis.

$$X—Y \xrightarrow{\text{heat or light}} X\cdot + \cdot Y$$

Another method is to introduce the odd electron from another species in the form of electron transfer *via* a redox reaction

$$Cu^{+} + X—Y \longrightarrow Cu^{2+} + X^{-} + \cdot Y$$

It is convenient to consider each method separately.

Thermolysis of covalent bonds

Clearly, the weaker the bond, the more readily it is broken. At sufficiently high temperatures (>700–800°C) *all* organic molecules break down to give radicals, but at reasonably convenient temperatures, *i.e.* <150°C, homolysis occurs at a suitable rate only in those molecules which contain relatively weak bonds (*ca* D = 160 kJ mol^{-1}, 40 kcal mol^{-1}, or less). A glance at the Appendix will reveal that peroxides, alkyl nitrites and certain azo compounds fall into this category, so it is not surprising that these compounds turn out to be widely used as radical initiators.

Thermolysis of peroxides

Dialkyl peroxides

Formation. These are usually prepared by formation of the alkyl hydroperoxide from hydrogen peroxide and the alcohol, followed by condensation with another molecule of the alcohol in the presence of *p*-toluenesulphonic acid, acting as an acidic catalyst:

$$ROH + H_2O_2 \longrightarrow ROOH + H_2O$$

$$RO\ddot{O}H \quad R—OH_2^{+} \longrightarrow ROOR + H_2O$$

Use. Low C-number primary and secondary alkyl peroxides are highly unstable and dangerous to handle, but tertiary alkyl peroxides such as t-butyl peroxide or α-cumenyl peroxide are more stable and decompose smoothly in solution at temperatures greater than 90°C to give the corresponding alkoxyl radicals. At higher temperatures (>130°C) further decomposition into methyl radicals and the corresponding ketones occurs:

$$Me_3COOCMe_3 \xrightarrow{100^\circ C} 2Me_3CO\cdot \xrightarrow{130^\circ C} Me\cdot + Me_2CO$$

$$PhCMe_2OOCMe_2Ph \xrightarrow{100^\circ C} 2PhCMe_2O\cdot \xrightarrow{130^\circ C} Me\cdot + PhCOMe$$

Di-t-butyl peroxide is a useful radical initiator because it is relatively safe to handle, being stable on storage, and gives radicals with localized single electrons with high reactivity towards most organic molecules. Its drawback is the relatively high temperature (*ca* 100°C) required for a convenient half life for thermolysis, *e.g.* $t_{\frac{1}{2}}$ at 100°C *ca* 1 week, $t_{\frac{1}{2}}$ at 120°C = 20 h.

Diacyl peroxides

Formation. These are readily prepared by reaction of acid chlorides with hydrogen peroxide (30 per cent) in a solvent such as carbon tetrachloride at 0°C in the presence of aqueous sodium hydroxide. The highly nucleophilic perhydroxyl ion reacts more readily than hydroxyl ion with the chloride to give the peracid anion, which in turn competes successfully for unchanged acid chloride to give the peroxide.

$$HOO^- + RCOCl \longrightarrow RCOOOH + Cl^-$$
$$RCOOO^- + RCOCl \longrightarrow RCOOOCOR + Cl^-$$

The reaction can be varied to give mixed peroxides by preisolation of the peracid followed by reaction with a different acid chloride; the pure peracid being best prepared, in this case, by methanolysis of a symmetrical diacyl peroxide with sodium methoxide.

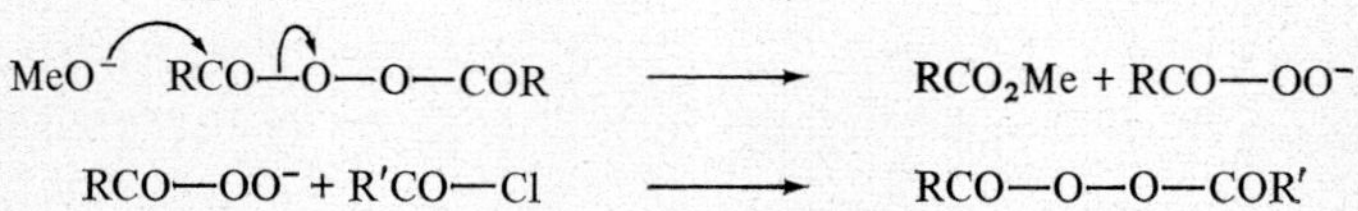

A very good method for the preparation of symmetrical peroxides using the condensing reagent dicyclohexylcarbodiimide has recently been reported:

$$2PhCO_2H + H_2O_2 + RN{=}C{=}NR \longrightarrow PhCOOOCOPh + RNHCONHR + H_2O$$

Use. Aliphatic diacyl peroxides, such as diacetyl peroxide, are highly unstable, even at room temperature, and are very dangerous to use. However, if they can be introduced into the reaction mixture without it blowing up they prove efficient initiators because they rapidly produce highly localized methyl radicals of high reactivity towards most substances. Long chain peroxides, such as didodecanoyl peroxide, are more stable and safe to handle and have industrial applications despite the resulting longer $t_{\frac{1}{2}}$ (decomposition).

$$\text{MeCOOOCOMe} \xrightarrow{25^\circ\text{C}} 2\text{MeCO}_2\cdot \longrightarrow \text{Me}\cdot + \text{CO}_2$$

Diaroyl peroxides, such as dibenzoyl peroxide, on the other hand, are much more satisfactory. They are reasonably stable on storage in the absence of heat, light, or friction,* they are crystalline solids capable of recrystallization with care at low temperatures, and decompose smoothly in solution at 80°C—half life *ca* 8 h in benzene—first to give benzoyloxyl radicals which, in most cases, decarboxylate at this temperature to give aryl radicals, *e.g.*

$$\text{Ph}\overset{\overset{\text{O}}{\|}}{\text{C}}\text{OO}\overset{\overset{\text{O}}{\|}}{\text{C}}\text{Ph} \xrightarrow{80^\circ\text{C}} 2\text{PhCO}_2\cdot \longrightarrow \text{Ph}\cdot + \text{CO}_2$$

$$(p\text{-NO}_2\text{C}_6\text{H}_4\text{CO}_2)_2 \xrightarrow{80^\circ\text{C}} p\text{-NO}_2\text{C}_6\text{H}_4\cdot + \text{CO}_2$$

Mixed aroyl alkanoyl peroxides, as might be expected, are intermediate in stability between symmetrical diaroyl- and dialkanoyl-peroxides. The activation energy for homolysis of the peroxide bond is lowered if the alkyl moiety can give rise to a delocalized radical. Thus, in the case shown below, simultaneous formation of CO_2, free phenyl and the thermodynamically more stable cinnamyl radical takes place easily.

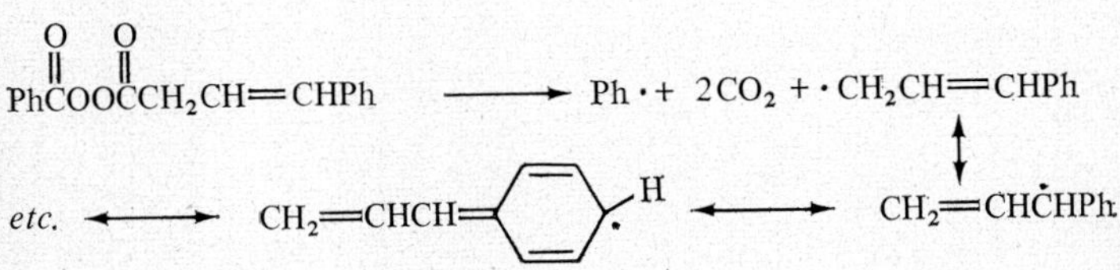

Peresters

Formation. Peresters are formed by reaction of alkyl hydroperoxides with acyl chlorides, in alkaline medium, *e.g.*

* Nb. Care should always be exercised in handling all peroxides.

$$PhCOCl + ROO^- \longrightarrow Ph\overset{\overset{O}{\|}}{C}OOR$$

Use. This combines the features of diacyl and dialkyl peroxides with no great advantage over either. The most commonly used is t-butyl perbenzoate,

$$Ph\overset{\overset{O}{\|}}{C}OOCMe_3 \longrightarrow Ph\cdot + CO_2 + Me_3CO\cdot$$

although peroxyoxalic esters are becoming increasingly useful as low temperature initiators, giving rise to alkoxyl radicals and two molecules of CO_2 smoothly at 40–50°C.

$$ROO\overset{\overset{O}{\|}}{C}—\overset{\overset{O}{\|}}{C}OOR \longrightarrow 2RO\cdot + 2CO_2$$

Peroxydicarbonates

Formation. Reaction of alkyl chloroformates with hydrogen peroxide in alkaline solution gives peroxydicarbonates.

$$RO\overset{\overset{O}{\|}}{C}Cl + HO_2^- \longrightarrow RO\overset{\overset{O}{\|}}{C}OO^-$$

$$RO\overset{\overset{O}{\|}}{C}OO^- + RO\overset{\overset{O}{\|}}{C}Cl \longrightarrow RO\overset{\overset{O}{\|}}{C}OO\overset{\overset{O}{\|}}{C}OR$$

Use. Peroxydicarbonates, such as di-isopropyl- or dicyclohexyl-peroxydicarbonate, are very useful initiators, decomposing smoothly at the usefully low temperature of 50°C to give the corresponding alkoxyl radicals.

$$RO\overset{\overset{O}{\|}}{C}OO\overset{\overset{O}{\|}}{C}OR \longrightarrow 2RO\cdot + 2CO_2$$

They are much used industrially, *e.g.* as initiators for the emulsion polymerization of vinyl chloride, where the disadvantage of low storage stability is often overcome by preparing them *in situ* in the polymerization vessel. Variations involving mixed peroxy sulphonyl carbonates are also used.

$$RO\overset{\overset{O}{\|}}{C}Cl + H_2O_2 + RSO_2Cl \longrightarrow$$

$$RO\overset{\overset{O}{\|}}{C}OOSO_2R \longrightarrow RO\cdot + CO_2 + RSO$$

Thermolysis of azo compounds

Thermolysis of azo compounds (RN=NR) to give two radicals and nitrogen does not occur at convenient temperatures (<150°C), despite the large driving force provided by the ejection of the stable molecule of nitrogen, unless there is also an additional energy contribution provided by delocalization in one or both of the possible radicals, or in rarer cases, by release of strain on decomposition of the azo compound.

Thus azobenzene (PhN=NPh) and azomethane (MeN=NMe) are stable under normal conditions of temperature. Arylazotriphenylmethanes and azobisalkylnitriles, on the other hand, decompose quite readily at 50°C and above.

Arylazotriphenylmethanes

Formation. These compounds are formed by reaction of triphenylmethyl chloride with arylhydrazine and subsequent oxidation of the resulting disubstituted hydrazine with mercuric oxide.

$$Ph_3CCl + ArNHNH_2 \longrightarrow Ph_3CNHNHAr$$
$$Ph_3CNHNHAr + HgO \longrightarrow Ph_3CN{=}NPh$$

Use. These compounds decompose smoothly, with almost simultaneous breaking of both bonds, at 50°C ($t_{\frac{1}{2}}$ *ca* 3 h) to give the heavily delocalized, and hence long lived, triphenylmethyl radical and the short lived aryl radical.

$$Ph_3CN{=}NPh \longrightarrow Ph_3C\cdot + PhN_2^{\cdot} \longrightarrow Ph\cdot + N_2$$

They are little used as initiators because the long lived triphenylmethyl radical tends to trap other radicals present and thus act as an inhibitor.

Azobisalkylnitriles

Formation. Condensation of ketones with HCN in the presence of hydrazine, followed by oxidation, readily yields azobisalkylnitriles.

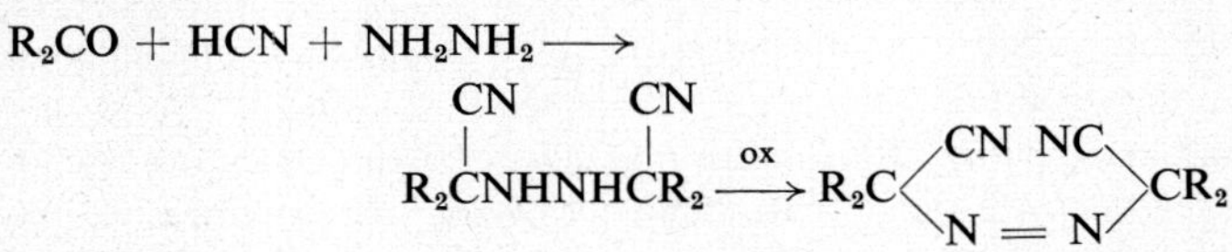

Use. The most widely used member of the series is azobisisobutyronitrile (AIBN), which decomposes readily at 50°C to give 2-cyanopropyl radicals.

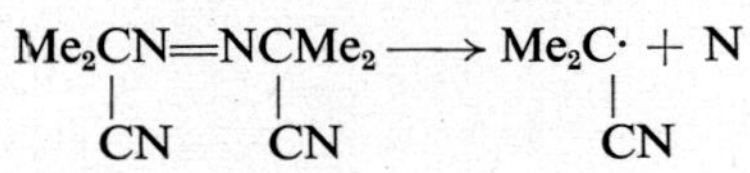

The efficiency of the initiator is marred by a significant loss, by recombination in the solvent cage, of the somewhat delocalized radicals,

$$Me_2\dot{C}CN \longleftrightarrow Me_2C{=}C{=}N\cdot$$

$$2Me_2\dot{C}CN \longrightarrow Me_2C(CN)C(CN)Me_2$$

but certain recently introduced derivatives of AIBN appear to be significantly better in this respect, *e.g.*

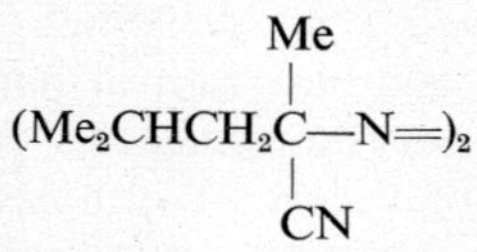 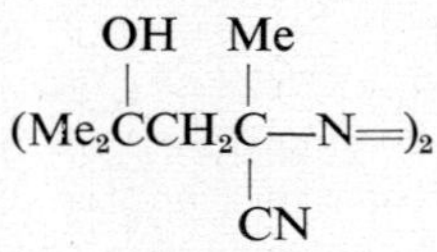

Dialkyl hyponitrites

Formation. These very easily decomposed, and hence potentially hazardous, esters of hyponitrous acid can be prepared by oxidation of alkylhydroxylamines or by reaction of silver hyponitrite with alkyl iodides.

$$2RONH_2 + Br_2/KOH \longrightarrow RON{=}NOR$$

$$Ag_2N_2O_2 + 2RI \longrightarrow RON{=}NOR$$

Use. They are used industrially as polymerization initiators and act by release of the relatively reactive alkoxyl radical at lower temperatures than those afforded by decomposition of the corresponding dialkyl peroxides.

$$RON{=}NOR \longrightarrow 2RO\cdot + N_2$$

Initiation via *diazonium compounds*

As described in Chapter 8, diazonium salts can be induced to give free aryl radicals, although such reactions have not been used extensively to initiate radical chain reactions.

The cage effect

Geminate recombination of initiator derived radicals. In many cases the efficiency of a radical initiator is decreased by recombination of the first formed radicals before reaction of the radicals with other molecules present can occur. Particularly good (or bad) examples of this are provided by the azobisalkylnitrile family discussed above and by peroxides such as diacetyl peroxide. The recombination reaction results from the very high absolute rates of coupling of radicals and the extent to which this instantaneous recombination occurs is a function of dilution by 'solvents', the nature of the solvent itself, the nature of the radicals, and hence the radical source.

In the gas phase, radicals produced by homolysis of the excited molecule of initiator can diffuse apart quite readily, thus reducing the chances of recombination. Even in solution, diffusion apart is rapid but much slower than in the gas phase. The time that the two radicals remain in close (reaction possible) proximity is of the order of 10^{-10} s and this is long enough for at least some of them to recombine, as a result of the high absolute rate of coupling. Such a process is known as a *geminate* reaction which is said to occur in the solvent *cage*. Once the geminate radical pair diffuses apart, of course, reaction occurs with 'solvent' molecules or by other routes, as outlined in Chapter 2. The extent of geminate recombination can be measured by the use of radical scavengers, such as iodine, butanethiol, or styrene, which react only with 'free' radicals. Thus, in the case of azobisisobutyronitrile, combination within the cage of the first formed $Me_2(CN)C\cdot$ radicals can cut down the efficiency of the initiation of polymerization of styrene to about 60 per cent—in benzene AIBN molarity $= 3 \times 10^{-3}$, styrene molarity $= 4.0$, at 60°C.

The questions of geminate combination and of the complex effect of the 'solvent' medium on radical–radical interactions are now being carefully investigated, as yet a complete explanation of the phenomenon has not emerged.

Photolysis of covalent bonds

Absorption of electromagnetic radiation can lead to homolysis of covalent bonds. The weaker the bond, the higher the wavelength of light needed. In theory, complete absorption of the quantum at 270 nm is equivalent to 450 kJ mol^{-1}, which is large enough to break most bonds (Appendix). In practice the energy is partially dissipated and the process is less efficient than is theoretically possible. Materials which are readily photolysed include alkyl nitrites and alkyl hypochlorites, both of which give alkoxyl radicals,

$$RONO \longrightarrow RO\cdot + NO\cdot$$
$$ROCl \longrightarrow RO\cdot + Cl\cdot$$

certain polyhalogenoalkanes (CF_3I, CCl_3Br), aryl iodides (ArI), and organomercury derivatives (RHgI) as well as the thermally labile peroxides and azo compounds mentioned in the preceding section. In most cases photolyses of these materials give two radical species, each of which may initiate a series of radical reactions, some of which are of considerable synthetical importance, *e.g.* the Barton reaction of substituted alkyl nitrites, *see* Chapter 8.

Photochemical production of radicals from carbonyl compounds has also received a lot of attention, mainly because they have an absorption band at about 320 nm, and thus can be photo-excited by light of this wavelength. Since this radiation can be easily produced *via* mercury arcs, and pyrex glassware is transparent to 320 nm radiation, the experimental procedure is relatively simple.

The first step in the photo reaction of carbonyl compounds is usually excitation to an excited state. This may be followed by decomposition to give radicals and/or neutral molecules: the radicals are then available to initiate radical chain reactions.

$$RCH_2CH_2CHO \xrightarrow{h\nu} RCH_2CH_2CHO^*$$

$$RCH_2CH_2CHO^* \longrightarrow \begin{cases} RCH{=}CH_2 + HCHO \\ RCH_2{-}CH_2\cdot + \dot{C}HO \\ RCH_2\cdot + \cdot CH_2CHO \end{cases}$$

Alternatively the photoexcited carbonyl compound may itself behave as a diradical; which in turn can initiate chain reactions, as in the photosensitized conversion of secondary alcohols into hydroperoxides in the presence of oxygen and a trace of benzophenone.

Initiation $Ph_2C{=}O \longrightarrow Ph_2\dot{C}{-}\dot{O}$

Propagation $Ph_2\dot{C}{-}\dot{O} + Me_2CHOH \longrightarrow Ph_2\dot{C}OH + Me_2\dot{C}OH$

$$Me_2\dot{C}OH + O_2 \longrightarrow Me_2C(OH)OO\cdot$$

$$Me_2CHOH + Me_2C(OH)OO\cdot \longrightarrow Me_2C(OH)OOH + Me_2\dot{C}OH \textit{ et seq}$$

Termination via radical coupling.

Redox reactions

Radicals can be generated by electron transfer, usually *via* metal ions of suitable redox potential, *e.g.* Cu^+, Ti^{3+}, Fe^{2+}, Co^{3+}. Typical examples are the reactions of diazonium salts with copper ions (Chapter 8) and the use of Fenton's reagent, which, besides being a good source of hydroxyl radicals, has the advantage of being

$$Fe^{2+} + H_2O_2 \longrightarrow Fe^{3+} + HO^- + HO\cdot$$

effective in aqueous solution, as has the corresponding reaction using Ti^{3+} ions. Such redox reactions are particularly important in commercial processes such as the low temperature emulsion polymerization of styrene by the initiator system cumenyl hydroperoxide and ferrous ion.

$$ROOH + Fe^{2+} \longrightarrow RO\cdot + HO^- + Fe^{3+}$$
$$RO\cdot + PhCH{=}CH_2 \longrightarrow \text{polymer}$$

Closely related to these are electrode reactions in which radicals are formed at the anode, *e.g.* in the Kolbe process (Chapter 8), or at the cathode, as in polarography.

$$RCO_2^- \xrightarrow{-1e} RCO_2\cdot \longrightarrow R\cdot + CO_2$$
$$ArN_2^+ \xrightarrow{+1e} ArN_2\cdot \longrightarrow Ar\cdot + N_2$$

The amine–peroxide system. Reactions between amines and peroxides are complex and often involve radicals. Tertiary amines, such as *NN*-dimethylaniline cause rapid acceleration of the decomposition of dibenzoyl peroxide to give radicals: a 0.01M solution of the peroxide in chloroform has a half life of many years at 20°C but the addition of 0.01M dimethylaniline reduces the half life to 13 min. Hence amines are often used as co-initiators with peroxides. The mechanism is believed to involve a redox process with the intermediate formation of an amine radical cation.

$$PhNMe_2 + (PhCO_2)_2 \longrightarrow PhCO_2^- \; PhN^+(Me)_2OCOPh$$

$$PhN^+(Me)_2OCOPh \longrightarrow [Ph\dot{N}Me_2]^+ + PhCO_2\cdot$$
$$PhN^+(Me)_2OCOPh \longrightarrow PhN^+(Me){=}CH_2 + PhCO_2H$$

Action of ionizing radiation. γ and other high energy radiations readily cause homolysis of covalent bonds to give radicals. The bonds are often broken in an indiscriminate manner, so this is not a widely used method of radical initiation. Biological damage following γ irradiation proceeds, at least in part, *via* free radical reactions.

5. Abstraction Reactions: Radical-Induced Halogenation and Autoxidation

Species such as the chlorine atom, alkyl-, aryl-, alkoxyl- or alkylperoxyl-radicals readily attack aliphatic C–H bonds by abstraction of a hydrogen atom and concomitant formation of an alkyl radical. The alkyl radical may then react in a variety of ways depending on circumstances.

$$\mathrm{X\cdot + R{-}H \longrightarrow XH + R\cdot \longrightarrow \text{further reaction}}$$
$$\mathrm{(X = Cl\cdot,\ Me\cdot,\ Ph\cdot,\ Me_3CO\cdot,\ R\cdot,\ RO_2^{\cdot}\ \textit{etc.})}$$

Since this general abstraction reaction is the first step of many processes of wide occurrence in organic chemistry and the biosphere, such as halogenation, oxidation, autooxidation, polymer degradation and smog formation, it is worth considering it in some detail.

Halogenation of C–H bonds

Direct halogenation. This is direct reaction of C–H bonds in alkanes or alkenes with halogens. The classic case is the chlorination of methane, which is very readily, and often explosively, achieved by exposing a mixture of the two gases to ultraviolet light. It is a *radical chain reaction* initiated by abstraction of a hydrogen atom by a chlorine atom to give hydrogen chloride and a methyl radical in an exothermic reaction (step *i*). This is followed by another exothermic reaction involving the first formed methyl radical and unreacted chlorine (step *ii*) which gives rise to methyl chloride and a chlorine atom, which then begins the chain all over again.

Initiation	$\mathrm{Cl \longrightarrow 2Cl\cdot}$	
Propagation	$\mathrm{Cl\cdot + CH_4 \longrightarrow HCl + \cdot Me}$	(*i*)
	$\mathrm{\cdot Me + Cl_2 \longrightarrow MeCl + Cl\cdot}$	(*ii*)

The characteristic of such a chain reaction is that in theory it requires only one molecule of chlorine to be homolyzed, reaction then proceeding either until one of the reactants is exhausted or until termination occurs *via* one of the following routes.

Termination	$\mathrm{2Cl\cdot \longrightarrow Cl}$
	$\mathrm{\cdot Me + Cl\cdot \longrightarrow MeCl}$
	$\mathrm{\cdot Me + \cdot Me \longrightarrow C_2H_6}$

In practice more than one initial fission is required, depending on how readily the termination reactions occur. In the case of the methane and chlorine reaction the propagating cycle, consisting of reactions (*i*) and (*ii*), can be repeated some 10^6 times before termination, the reaction then being said to have a *chain length* of 10^6.

So far, replacement of only one hydrogen by chlorine has been considered; in practice, of course, the first formed methyl chloride also enters into further radical chain reactions with chlorine to give, ultimately, a mixture of CH_2Cl_2, $CHCl_3$ and CCl_4, thus:

$$Cl\cdot + MeCl \longrightarrow HCl + \cdot CH_2Cl$$
$$\cdot CH_2Cl + Cl_2 \longrightarrow CH_2Cl_2 + Cl\cdot \textit{ et seq.}$$

Bromination of methane under the influence of ultraviolet radiation occurs much more slowly than chlorination because the abstraction step in bromination is endothermic. That any reaction occurs is due to the compensating exothermicity of the second propagating step:

$$Br_2 \longrightarrow 2Br\cdot$$
$$Br\cdot + CH_4 \longrightarrow HBr + \cdot Me \qquad \Delta H = +54.5 \text{ kJ mol}^{-1}$$
$$\cdot Me + Br_2 \longrightarrow MeBr + Br\cdot \qquad \Delta H = -109 \text{ kJ mol}^{-1}$$

Fluorination is very rapid, even more so than chlorination, because both steps are highly exothermic and, moreover, the bond dissociation energy of fluorine is very low. In fact the bond in F_2 is one of the weakest in a stable molecule while that in H–F is one of the strongest.

Turning from methane, which has only one type of C–H bond, the halogenation of materials containing different types of C–H bond will be considered. In alkanes the relative ease of abstraction of hydrogen by halogen atoms parallels the relative strengths of the C–H bonds, the order of reactivity being tertiary > secondary > primary, and hence chlorination of 2-methylpropane gives the following approximate product distribution:

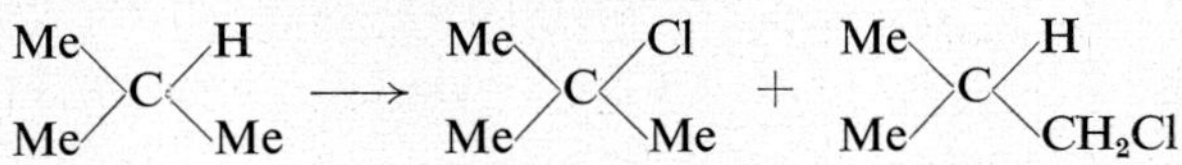

35 per cent 65 per cent

Thus, allowing for the statistical factor of there being nine equivalent primary hydrogen atoms to one tertiary, it can be calculated that chlorine atoms attack the latter 4.8 faster than the former. On the other hand, bromination of 2-methylpropane gives exclusive bromination at the tertiary hydrogen atom. This result still agrees with

$$Me_3CH \longrightarrow Me_3CBr \quad \text{100 per cent}$$

the qualitative picture that a tertiary C–H bond is more readily attacked than a primary analogue, but quantitatively it is quite different from the pattern of chlorination. Clearly chlorination is much less *selective* than bromination.

In general *selectivity* is a function of the extent of bond-breaking or making contributions in the transition state. If the species is highly reactive in the sense that its reactions are exothermic as is the case with a chlorine atom,

$$Cl\cdot + H{-}R \longrightarrow Cl\cdots\cdots H\cdots R \longrightarrow HCl + R\cdot$$

the transition state resembles the reactants (rather than the products) and the nature of the C–H bond would therefore have little effect on the rate of reaction, *i.e.* the highly reactive chlorine atom (in chlorination of RH) is non selective (*Fig. 2A*). In bromination, where the abstraction step is endothermic (*see* below), involving a relatively low energy species, the activation energy is high, the transition state resembles the products, and hence the nature of the R–H bond becomes predominant, leading to higher selectivity (*Fig. 2B*).

$$Br\cdot + H{-}R \longrightarrow Br\cdots H\cdots\cdots R \longrightarrow HBr + R\cdot$$

Allylic halogenation

Halogens react under ionic conditions with alkenes to give 1,2-dihaloalkanes, but under radical conditions, *e.g.* Cl_2 at 500°C, reaction takes place *via* hydrogen abstraction at the weak allylic or α-methylenic C–H bond; a reaction favoured by the relatively high thermodynamic stability of the resulting delocalized allyl-type radical. The alternative reaction, of addition of a chlorine atom to the double bond, is reversible and the equilibrium is unfavourable at this temperature.

FIG. 2. Selectivity of species.

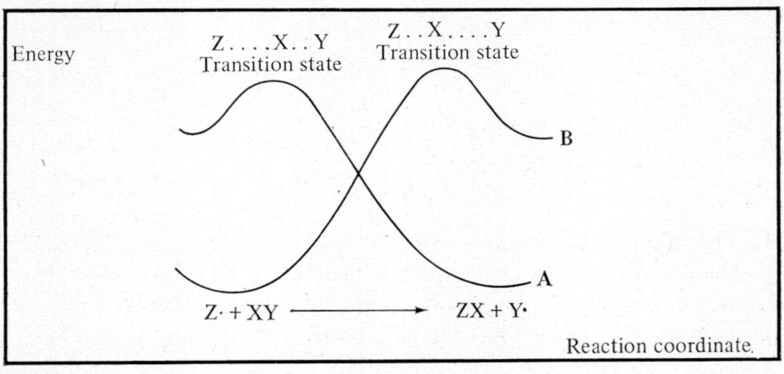

$$Cl\cdot + RCH_2CH{=}CH_2 \longrightarrow HCl + R\dot{C}HCH{=}CH_2$$

$$\rightleftharpoons \qquad\qquad\qquad \updownarrow$$

$$RCH_2\dot{C}HCH_2Cl \qquad RCH{=}CHCH_2Cl \longleftarrow RCH{=}CH\dot{C}H_2$$

This reaction is particularly favoured in the case of cyclohexene, which gives 3-chlorocyclohexene on high temperature chlorination.

Polar effects in chlorination. Whereas in most radical reactions polar effects are very small, in chlorination of alkane derivatives polar effects are of considerable importance. The direction of these effects indicates that the chlorine atom has an electrophilic component in its radical behaviour. Thus, Table 1 shows that substitution of the more electron donating groups R, in $RC(CH_3)_3$, increases the rate of chlorination relative to those cases where R is electron withdrawing.

Table 1. Effect of substituents on chlorination

$$RC(Me)_2{-}Me + Cl\cdot \longrightarrow RC(Me)_2{-}\dot{C}H_2 + HCl$$

R	Relative rate
Me_3C	1.2
p-$NO_2C_6H_4$	0.5
C_6H_5CO	0.4
ClCO	0.19
CN	0.17

Bromination via N-*bromosuccinimide.* *N*-Bromosuccinimide (NBS) in the presence of a radical initiator (light or peroxides) very readily brominates at weak C–H bonds. The reaction works particularly well, therefore, at allylic positions, *e.g.*

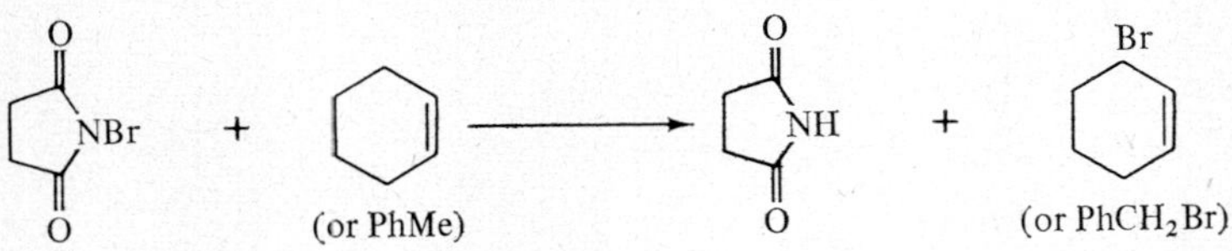

The mechanism of the reaction has only recently been established (Tedder) and it is now clear that the brominating agent is molecular bromine; present in and maintained at, low concentrations by reaction of NBS with the HBr produced in the reaction

$$Br_2 \longrightarrow 2Br\cdot \quad \text{(initiation)}$$

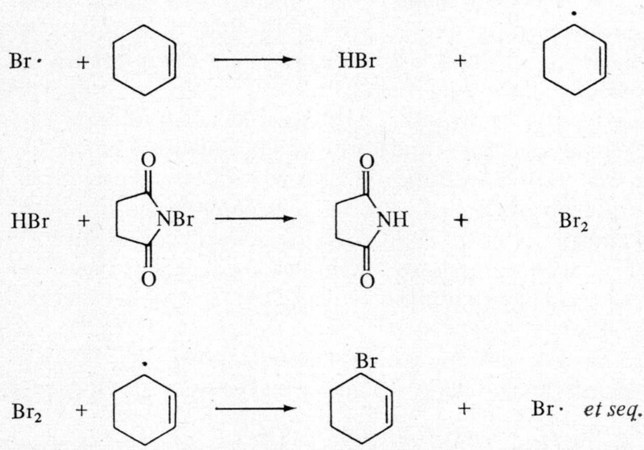

Since the competing addition of bromine atoms to the double bond is reversible, the presence of a low concentration of bromine, produced in this way, favours the allylic substitution.

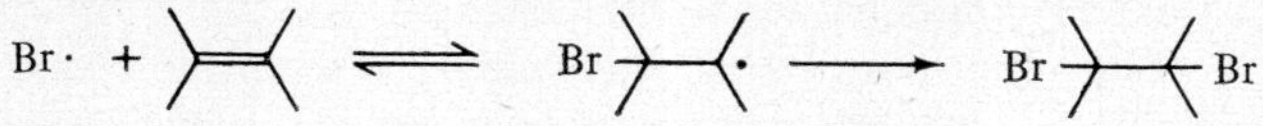

The principles of selectivity which apply to direct halogenation by chlorine or bromine outlined above also apply in the case of *N*-halosuccinimides. Thus *N*-bromosuccinimide is much more selective, and hence more useful synthetically, than chlorine in halogenation of C–H bonds.

The photonitrosation process for the production of nylon 6. This commercial process is based on abstraction of hydrogen by halogen atoms. It is a non chain reaction and consists of photolysis of nitrosyl chloride in the presence of cyclohexane (RH), when the following sequence occurs

$$NOCl \longrightarrow NO + Cl\cdot$$

$$Cl\cdot + RH \longrightarrow R\cdot + HCl$$

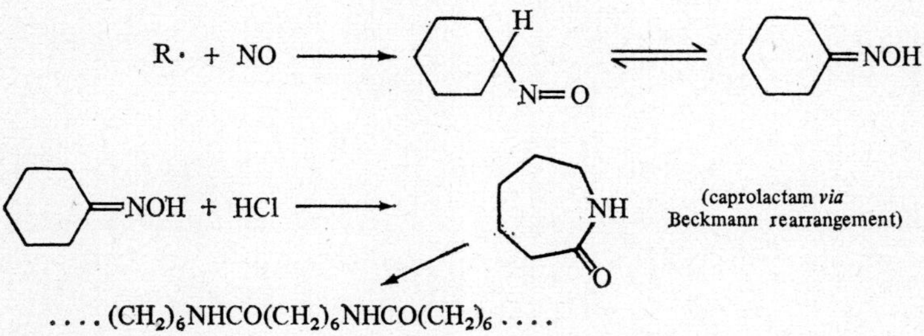

$$\ldots.(CH_2)_6NHCO(CH_2)_6NHCO(CH_2)_6\ldots.$$

Autoxidation

Autoxidation is the low temperature oxidation of organic compounds by oxygen, unaccompanied by the production of flame. Since oxygen is a diradical it is not surprising to find that autoxidations are radical processes, and are acclerated by the presence of radical initiators or ultraviolet light.

Degradation of polymers. Autoxidative degradation of plastics, rubber, paints, lacquers and fibres which lead ultimately to breakdown of the material is an important process in two senses: first because it is important to retard it in many circumstances; it is undesirable, for example, to have rapid, air-catalyzed decomposition of motor car tyres. Second, because sometimes the acceleration of decomposition *via* aerial autoxidation is desirable, *e.g.* in the degradation of plastic litter.

The most common autoxidation involves the formation of a hydroperoxide in a chain reaction involving an alkylperoxyl radical.

$$\text{Initiator} \longrightarrow \text{R}\cdot$$
$$\text{R}\cdot + \text{O}_2 \longrightarrow \text{ROO}\cdot$$
$$\text{ROO}\cdot + \text{RH} \longrightarrow \text{ROOH} + \text{R}\cdot$$

In this case RH could be a polymer such as polythene or a polyisoprene rubber, the latter being particularly susceptible because of its weak allylic C–H bonds. The hydroperoxide so formed can then undergo a further series of decompositions leading to degradation of the material. Obviously, the mechanism and products will depend on the structure of the polymer but in each case the picture is complex.

Polypropylene

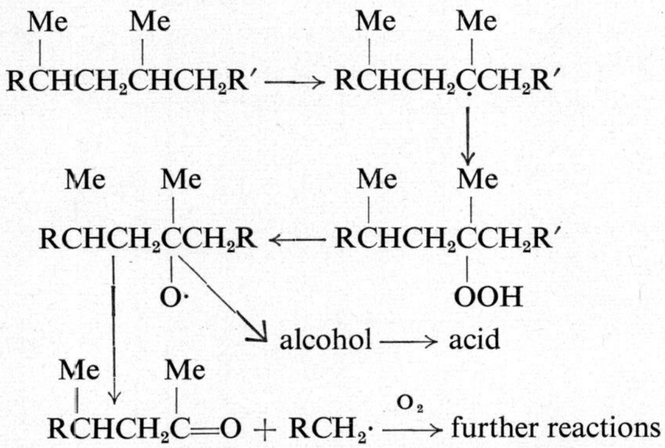

Since these reactions proceed by a chain mechanism it is possible to stabilise the systems by use of additives which are either long lived free radicals or are materials which readily give such radicals. These then stop the radical chains from getting under way, *e.g.*

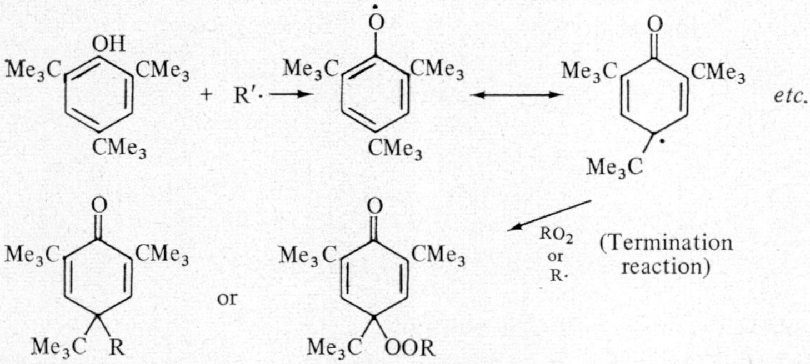

Alternatively, it is possible to accelerate radical decomposition by the addition of radical initiators, or by designing polymers with inbuilt points of high reactivity towards radicals, and much work along these lines is now in progress.

Production of phenol. Autoxidation is used commercially in the production of phenol and acetone from isopropylbenzene (cumene). The first formed product is cumene hydroperoxide which, under the influence of acid, undergoes hydrolysis and rearrangement as shown:

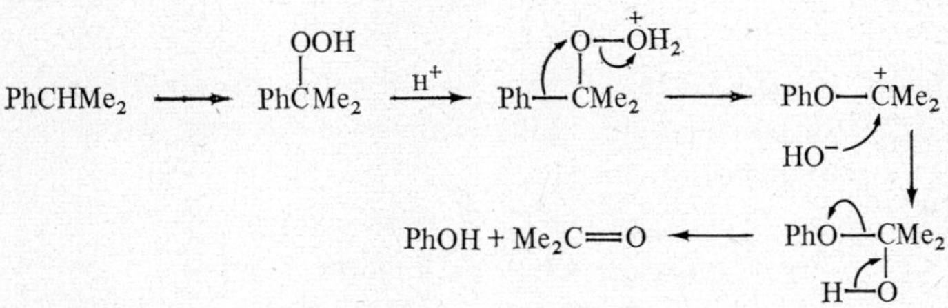

Autoxidation of ethers. Ethers peroxidize very readily in air and the resulting peroxides make the distillation of ethers to dryness a very hazardous operation. The first step is autoxidation to give an α-alkoxyhydroperoxide, which then undergoes successive condensations to give a highly explosive dialkyl peroxide.

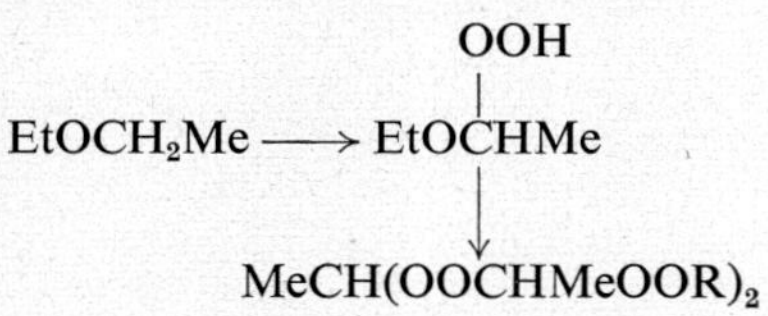

Thus diethyl ether should be washed with aqueous ferrous sulphate or passed through activated alumina before use.

Autoxidation of aldehydes. This occurs very readily and leads to the corresponding carboxylic acid by the following chain reaction.

$$\mathrm{RCHO} \xrightarrow{\mathrm{R\cdot}} \mathrm{R\dot{C}O} \xrightarrow{\mathrm{O_2}} \mathrm{RC({=}O)OO\cdot}$$

$$\mathrm{RC({=}O)O_2\cdot} + \mathrm{RCHO} \longrightarrow \mathrm{RC({=}O)O{-}OH} + \mathrm{R\dot{C}O}$$

$$\mathrm{RCOOOH} + \mathrm{RCHO} \longrightarrow \mathrm{2RCO_2H}$$

Reaction with C–halogen bonds

Weak C-halogen bonds (Appendix) are readily attacked by radicals. Typical examples are:

$$\mathrm{Ph\cdot} + \mathrm{CCl_4} \longrightarrow \mathrm{PhCl} + \mathrm{\cdot CCl_3}$$
$$\mathrm{Ph\cdot} + \mathrm{BrCCl_3} \longrightarrow \mathrm{PhBr} + \mathrm{\cdot CCl_3}$$
$$\mathrm{2\dot{C}Cl_3} \longrightarrow \mathrm{C_2Cl_6}$$

The ease with which bromotrichloromethane undergoes bromine abstractions is used in a simple radical chain bromination of benzylic C–H bonds:

$$\mathrm{R\cdot} + \mathrm{BrCCl_3} \longrightarrow \mathrm{RBr} + \mathrm{\cdot CCl_3}$$
$$\mathrm{\cdot CCl_3} + \mathrm{PhCH_3} \longrightarrow \mathrm{CHCl_3} + \mathrm{PhCH_2\cdot}$$
$$\mathrm{PhCH_2\cdot} + \mathrm{BrCCl_3} \longrightarrow \mathrm{PhCH_2Br} + \mathrm{\cdot CCl_3} \quad \textit{et seq.}$$

Closely related to this is the use of tributyltin hydride in the radical chain conversion of alkyl halides to alkanes, a reaction of some importance in synthesis.

Initiation $$\mathrm{Bu_3SnH} + \mathrm{R\cdot'} \longrightarrow \mathrm{R'H} + \mathrm{Bu_3Sn\cdot}$$
$$\mathrm{Bu_3Sn\cdot} + \mathrm{RBr} \longrightarrow \mathrm{Bu_3SnBr} + \mathrm{R\cdot}$$
$$\mathrm{R\cdot} + \mathrm{Bu_3SnH} \longrightarrow \mathrm{RH} + \mathrm{Bu_3Sn\cdot} \quad \textit{et seq.}$$

6. Addition of Radicals to Aliphatic Unsaturated Systems

The first recognized example of radical induced addition reactions of alkene double bonds occurred in 1933 when Kharasch and Mayo noted that the addition of hydrogen bromide to allyl bromide proceeded 'abnormally' in the presence of added peroxides or light, resulting in orientation of addition opposite to that produced with the pure reactants alone. This phenomenon, 'the peroxide effect', was later explained simultaneously by Kharasch in the US and by Hey and Waters in this country in terms of a free radical initiated addition reaction.

Initiation

$$\text{Peroxide} \longrightarrow \text{R}\cdot$$
$$\text{R}\cdot + \text{HBr} \longrightarrow \text{RH} + \text{Br}\cdot$$
$$\text{Light} + \text{HBr} \longrightarrow \text{H}\cdot + \text{Br}\cdot$$
$$\text{Br}\cdot + \text{CH}_2\text{=CHCH}_2\text{Br} \longrightarrow \text{BrCH}_2\dot{\text{C}}\text{HCH}_2\text{Br}$$

Propagation

$$\text{BrCH}_2\dot{\text{C}}\text{HCH}_2\text{Br} + \text{HBr} \longrightarrow \text{BrCH}_2\text{CH}_2\text{CH}_2\text{Br} + \text{Br}\cdot$$
$$\text{Br}\cdot + \text{CH}_2\text{=CHCH}_2\text{Br} \longrightarrow \text{BrCH}_2\dot{\text{C}}\text{HCH}_2\text{Br} \textit{ etc.}$$

Termination. Any dimerization, coupling or disproportionation reaction involving the participating radicals.

This reaction was in direct contrast to the ionic counterpart, which occurred with pure components in the absence of peroxides or light, and which proceeded *via* the more stable of the two possible carbonium ions.

$$\text{CH}_2\text{=CHCH}_2\text{Br} + \text{H}^+ \longrightarrow \text{Me}\overset{+}{\text{C}}\text{HCH}_2\text{Br}$$
$$\text{Me}\overset{+}{\text{C}}\text{HCH}_2\text{Br} + \text{Br}^- \longrightarrow \text{MeCHBrCH}_2\text{Br}$$

An important feature of the radical reaction is that it is a *chain* reaction. In theory it requires only trace amounts of radical initiator to start the reaction; once initiated the chain regeneration of the bromine atom sustains the reaction until either all the alkene or all the hydrogen bromide is consumed. The reaction then terminates by radical–radical interaction. In practice, radical–radical interaction competes with the chain propagation reaction to a significant extent, depending on the system, and so cuts down the effectiveness of the initiator so that rather more than a trace (up to 0.01 mole per mole of alkene) of peroxide is necessary for a good conversion, in this case.

This reaction, once of theoretical interest only, has now become the prototype for a very large number of radical induced addition reactions of molecules to alkenes; reactions which in many instances do not occur at all in the absence of radical initiation by light, peroxides or related initiators. It is also important because the initiation step is common to the first step of radical induced polymerization.

Before discussing the scope of this radical chain addition reaction, however, it is helpful to consider the possible reactions which can occur in the general case of addition of a molecule XY to an unsymmetrical alkene.

Radical addition of XY to $RCH{=}CH_2$

Initiation can occur by photolysis of XY or by reaction of XY with a radical P·—produced by thermolysis of a peroxide—present in catalytic amount, to give Y·. Addition of Y· to the unsymmetrical alkene, $RCH{=}CH_2$, proceeds *via* the more thermodynamically stable of the two possible isomeric intermediate radicals, *i.e.* the secondary radical $R\dot{C}HCH_2Y$ rather than the primary radical $RCHYCH_2\cdot$. This preference can be rationalized in terms of the greater possible extent of delocalization of the hyperconjugative type in the secondary radical.

$$R\dot{C}HCH_2Y \longleftrightarrow RCH{=}CHY\overset{\cdot H}{} \text{ (two forms)}$$

$$RCHYCH_2\cdot \longleftrightarrow RCY{=}CH_2\overset{\cdot H}{} \text{ (one form)}$$

Initiation

$$XY \longrightarrow X\cdot + Y\cdot$$

$$\text{Peroxide} \longrightarrow P\cdot$$

$$P\cdot + XY \longrightarrow PX + Y\cdot$$

$$Y\cdot + RCH{=}CH_2 \longrightarrow R\dot{C}HCH_2Y$$

Propagation

$$XY + R\dot{C}HCH_2Y \longrightarrow Y\cdot + RCHXCH_2Y \qquad (i)$$

[chain transfer to XY giving a 1 : 1 adduct]

$$R\dot{C}HCH_2Y + RCH{=}CH_2 \longrightarrow R\dot{C}HCH_2CHRCH_2Y$$

$$\longrightarrow R\dot{C}H(CH_2CHR)_nCH_2Y \qquad (ii)$$

$$XY + R\dot{C}H(CH_2CHR)_nCH_2Y \longrightarrow Y\cdot + RCHX(CH_2CHR)_nCH_2Y$$

[chain transfer to XY giving a telomer (*n* : 1 adduct)]

Termination

$$Y\cdot + Y\cdot \longrightarrow Y{-}Y$$

$$Y\cdot + R\dot{C}HCH_2Y \longrightarrow RCHYCH_2Y \text{ etc.}$$

The effect is much more marked if we take the nature of R into account, *e.g.* in the case of $R'CH_2CH{=}CH_2$

$$R'CH_2\dot{C}HCH_2Y \longleftrightarrow \underset{\text{(two forms)}}{\overset{H\cdot}{R'CH{=}CHCH_2Y}} \longleftrightarrow \underset{\text{(two forms}}{\overset{H\cdot}{R'CH_2CH{=}CHY}}$$

$$R'CH_2CHYCH_2\cdot \longleftrightarrow \underset{\text{(one form only)}}{\overset{H\cdot}{R'CH_2CY{=}CH_2}}$$

The secondary radical, $R\dot{C}HCH_2Y$, formed can now react with another molecule of XY, in a so-called *chain transfer step*, to give the 1 : 1 adduct $RCHXCH_2Y$ and a new radical Y· which can continue the chain by adding to another molecule of alkene. Alternatively, the secondary radical $R\dot{C}HCH_2Y$ can itself undergo the same type of radical addition to a molecule of alkene to give a new secondary radical $R\dot{C}HCH_2CHRCH_2Y$. Similar, consecutive reactions between the growing radical and successive molecules of alkene to give a high molecular weight radical, $R\dot{C}H(CH_2CHR)_nCH_2Y$, are therefore possible. Chain transfer between this radical and XY leads to an *n* : 1 adduct usually called a *telomer*.* The reaction ceases when either all XY or all of the alkene are consumed, termination then occurring by radical–radical interactions (coupling or disproportionation) of the remaining radicals present. It should be noted that the reaction sequence requires that the total amount of products (PX) derived from the peroxide initiator must equal the total amount of termination derived products, *e.g.* Y–Y *etc.* The ratio of the rate constants of reactions (*i*) and (*ii*) is known as the *chain transfer constant* for the overall process involving the particular pair of reactants XY and $RCH_2CH{=}CH_2$. In practice the chain transfer constant is obtained by noting the extent of the lowering of polymerization (by mw measurement) of a polymerizable monomer, such as styrene, by added chain transfer reagents XY. This gives an invaluable quantitative guide to the ease of abstractability of atoms from different substrates. In general, butane-1-thiol (BuS–H) and Br–CCl_3 which have easily broken bonds, have high chain transfer constants while alcohols ($HOCH_2$–H) have low constants.

Returning to the general addition process, in theory the addition reaction is capable of giving a whole range of adducts consisting of one molecule of XY with increasing numbers of alkene molecules. In the case of XY, where X = H, the 1 : 1 adduct, RCH_2CH_2Y, corresponds to the alkylation of HY by the alkene, obviously a

* As opposed to a polymer (polymonomer) which contains elements of the monomer $RCH = CH_2$, only.

reaction of some general synthetic interest, depending on the nature of Y. On the other hand, high molecular weight materials derived from alkenes are also of interest as possible new plastic materials. All that is necessary is to use our knowledge of radical reactions to divert the reaction in the required direction.

Factors favouring the formation of 1:1 adducts rather than polymers (telomers)

Simple 1 : 1 adducts are favoured by (*a*) High concentration of XY. (*b*) Low concentration of $RCH{=}CH_2$. (*c*) A weak X–Y bond. (*d*) A low tendency to polymerization of $RCH{=}CH_2$. These factors all tend to ensure that the probability of occurrence of reaction (*i*) is greater than that of reaction (*ii*), but with the present state of knowledge it is impossible to predict the exact course of a given reaction in the absence of precise data on the energetics of each of the possible reaction steps available. Clearly, in order that a significant kinetic chain length be achieved, addition of the radical $Y\cdot$ to the double bond ($Y\cdot + RCH{=}CH_2 \rightarrow R\dot{C}HCH_2Y$) and the transfer step involving attack of the intermediate radical $R\dot{C}HCH_2Y$ on XY must be fast compared with all possible termination steps; *i.e.* both of these reactions should preferably be exothermic otherwise radical chain initiation is impossible, no matter how efficient an initiator is used. Even then, too high an energy of activation or too low an entropy factor may rule out the possibility of successful reaction. (Energies of the relevant steps in a few selected addition reactions are given in Table 2.)

Table 2. Approximate energies of propagation steps in radical addition to propene

X–Y	Addition $\cdot Y + RCH{=}CH_2 \rightarrow R\dot{C}HCH_2Y$ ΔH kJ mol^{-1} (kcal mol^{-1})	Displacement $R\dot{C}H_2CH_2Y + XY \rightarrow RCHXCH_2Y + Y\cdot$ ΔH kJ mol^{-1} (kcal mol^{-1})
H–Cl	−125 (−30)	38 (9)
H–Br	−38 (−9)	−29 (−7)
H–I	12.5 (3)	−96 (−23)
H–OH	−150 (−36)	109 (26)
Cl–CCl_3	−75 (−18)	−17 (−4)

Thus, the addition of HBr and carbon tetrachloride proceeds smoothly since both steps are exothermic in each case. HCl and water do not add because the displacements steps are endothermic, largely due to the high strength of the H–Cl and H–OH bonds, whereas H–I does not add because the addition step is endothermic and reversible, as a result of the weak C–I bond. This factor is used in the radical initiated isomerization of *cis*- to *trans*-alkenes

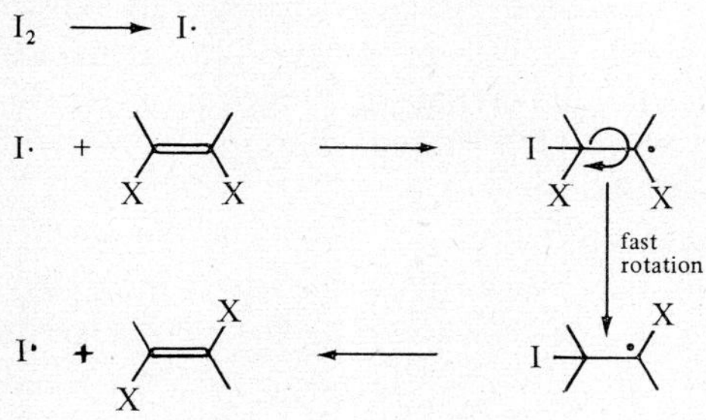

by iodine, because rotation in the intermediate radical takes place before the iodine atom has time to leave in the reversible step.

Radical addition of polyhalogenomethanes to olefins. The radical additions of carbon tetrachloride, which has a relatively strong C–Cl bond, and bromotrichloromethane, which has a weak C–Br bond, to oct-1-ene and to styrene illustrate the factors discussed above.

Reaction of oct-1-ene (1 mole), carbon tetrachloride (5 mole) and dibenzoyl peroxide (0.01 mole) gives a mixture of the 1 : 1 adduct (mainly) and significant amounts of the 2 : 1 and 3 : 1 adducts, whereas reaction of styrene (1 mole) even in a very large excess of carbon tetrachloride (50 mole) always gives high molecular weight, telomeric products. These observations reflect a combination of factors, including the strong C–Cl bond in carbon tetrachloride and the delocalization of the initiating intermediate benzyl-type radical.

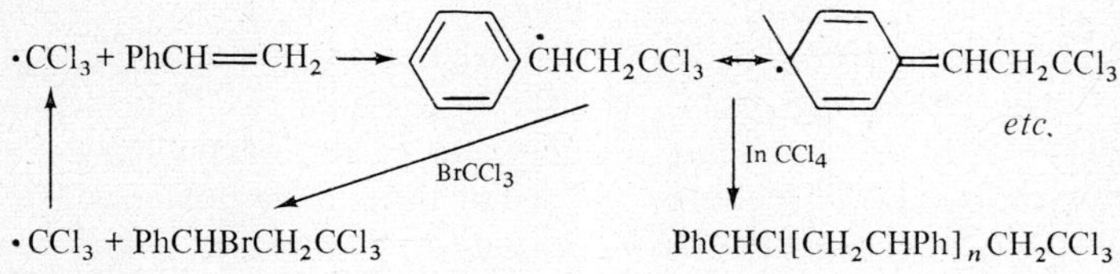

Use of bromotrichloromethane (3 moles) in the latter case instead of carbon tetrachloride leads to a very high yield of the 1 : 1 adduct.

Radical alkylation reactions

Fair to good yields of the corresponding 1 : 1-adducts can be obtained on radical induced reaction of non-conjugated alkenes with alcohols, malonic ester, acetoacetic ester, acetic acid, and related compounds, which must be present in large excess (20–100 moles) because strong C–H bonds have to be broken. In the case of alcohols, the more easily abstracted tertiary hydrogen atom in isopropanol leads to a higher yield of 1 : 1 adduct than in the case of ethanol under comparable conditions:

Alkylation of alcohols

$$R\cdot + Me_2CHOH \longrightarrow RH + Me_2\dot{C}OH$$

$$Me_2\dot{C}OH + R'CH{=}CH_2 \longrightarrow R'\dot{C}HCH_2C(OH)Me_2$$

$$R'\dot{C}HCH_2C(OH)Me_2 \xrightarrow{Me_2CHOH} Me_2\dot{C}OH + R'CH_2CH_2C(OH)Me_2$$

Alkylation of acetic acid and derivatives

$$R\cdot + MeCO_2H \longrightarrow RH + \cdot CH_2CO_2H$$

$$\cdot CH_2CO_2H + R'CH{=}CH_2 \longrightarrow R'\dot{C}HCH_2CH_2CO_2H$$

$$R'\dot{C}HCH_2CH_2CO_2H \xrightarrow{MeCO_2H} \dot{C}H_2CO_2H + R'CH_2CH_2CH_2CO_2H$$

Similarly

$$CH_2(CO_2Et)_2 \longrightarrow \dot{C}H(CO_2Et)_2 \xrightarrow{RCH=CH_2} RCH_2CH_2CH(CO_2Et)_2$$

$$MeCOCH_2CO_2Et \longrightarrow MeCO\dot{C}HCO_2Et \longrightarrow RCH_2CH_2CH(COMe)(CO_2Et)$$

The last three reactions are particularly useful in synthesis since direct alkylation of acetic acid is very difficult under ionic conditions, while that of malonic ester is the well known malonic ester synthesis.

Radical induced polymerization and telomerization

Reactions of radicals with conjugated olefins give rise to delocalized radicals which are less reactive than those obtained from simple, non conjugated alkenes. These delocalized radicals tend to react by successive addition to further molecules of olefin because this is the lower energy pathway available in these cases. Hence homopolymers of these olefins are formed very easily, particularly if the olefinic monomer is in excess. Such reactions have great industrial significance and are the basis of the formation of polystyrene, polyacrylonitrile, polymethyl methacrylate *etc.* from styrene (vinylbenzene, $PhCH{=}CH_2$), acrylonitrile (vinyl cyanide, $CH_2{=}CHCN$), methyl methacrylate (methyl 2-methylprop-2-enoate, $CH_2{=}C(CH_3)$-CO_2CH_3) respectively.

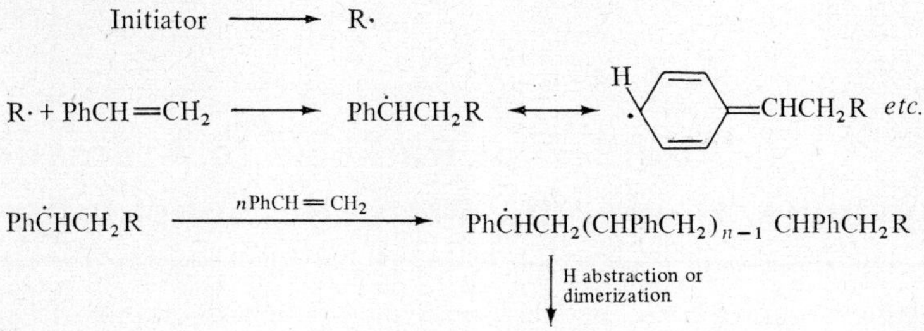

polystyrene

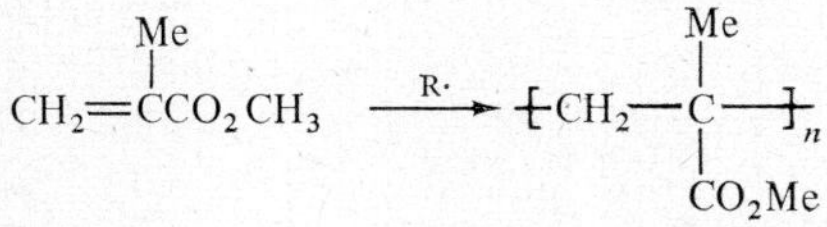

polymethylmethacrylate 'perspex'

Similar polymerization of simple alkenes such as ethylene or propene are less easily achieved, but nevertheless the use of high pressure and temperature with a suitable initiator enables ethylene to be converted into polythene, perfluoroethylene into polyperfluoroethylene (teflon) and vinyl chloride into polyvinyl chloride (pvc). In addition to polymers derived solely from one olefin, or monomer, many useful plastics are produced by allowing two or more monomers to polymerize together to give so-called copolymers. Variation of the types and concentration of the monomers leads to a wide variety of plastics of different physical and chemical properties. One of the best examples of this is provided by the production of styrene–butadiene rubber (sbr) which is now more than 2m long tons *per annum*.

Inhibition of polymerization. It is often desirable to prevent polymerization, *e.g.* in the storage of styrene monomer, while awaiting subsequent polymerization. This can be achieved by the presence of relatively small amounts of a radical scavenger such as 2,4,6-tri-t-butylphenol (*see* Chapter 3), which, on reaction with any stray radical initiator, will give a radical of long life which, in turn, will react with further radicals in a chain ending, radical coupling reaction.

Radical addition to alkynes

Additions to alkynes are similar to those involving alkenes, but are less well known. The first step is the formation of an alkene which then may react further to give a fully saturated product.

$$EtSH + PhC{\equiv}CH \rightarrow PhCH{=}CHSEt \rightarrow PhCH_2CH(SEt)_2$$

7. Addition of Radicals to Aromatic Systems: Homolytic Aromatic Substitution

Aromatic substitution, whether it involves radicals or ions, is an addition–elimination process. The first step involves addition to an aromatic double bond, thus disrupting the aromatic system.

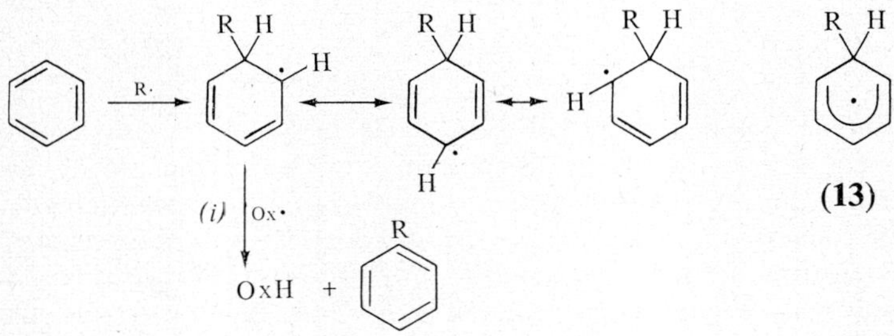

(13)

The intermediate cyclohexadienyl radical, which is delocalized and is often written as (**13**), can revert to the stable aromatic form by eliminating a hydrogen atom. This does not occur spontaneously, but takes place readily in the presence of an oxidizing agent, which may be another radical, a metal ion, a peroxide *etc.* This is in contrast to addition to a simple alkene double bond, where there is no corresponding energy bonus on reverting to the unsaturated

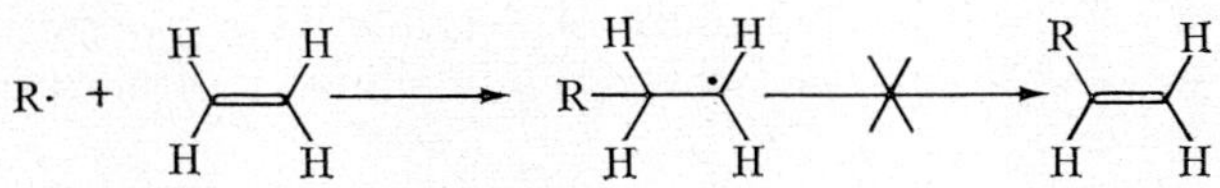

form. The radical substitution reaction most studied is phenylation. This is commonly effected by dibenzoyl peroxide, when the reaction in benzene takes place mainly as indicated opposite:

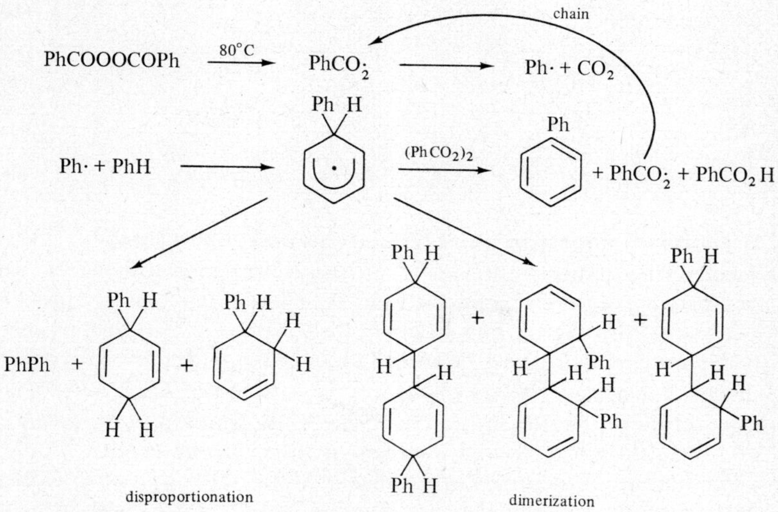

Thus, in addition to biphenyl and benzoic acid, products of dimerization and disproportionation of the partially delocalized intermediate cyclohexadienyl radical are formed—about 50 per cent—depending on the concentration of the reactants.

Other phenylating agents may also be used, *e.g.* diazonium salts, *N*-nitrosoacetanilide, phenylazotriphenylmethane, but in each case the side products and nature of the oxidative step (*i*) are different (Chapter 8).

The reaction may be extended by using substituted phenyl radicals, *e.g.* *p*-chlorophenyl radicals formed by decomposition of bis-(*p*-chlorobenzoyl) peroxide. This radical is a highly localized radical in the sense that the single electron is in an sp^2 orbital and *remains in it* (*see* **14**), delocalization into the orthogonal π-cloud being impossible, since overlap is impossible, hence reaction of the *p*-chlorophenyl radical with benzene gives only 4-chlorobiphenyl and not any of its isomers.

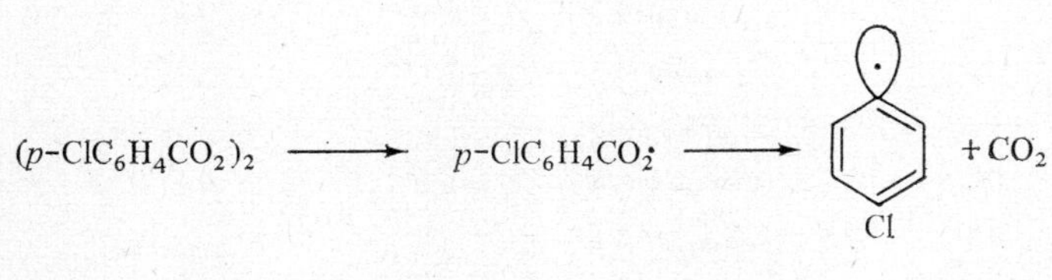

(14) 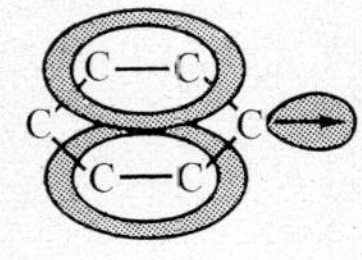

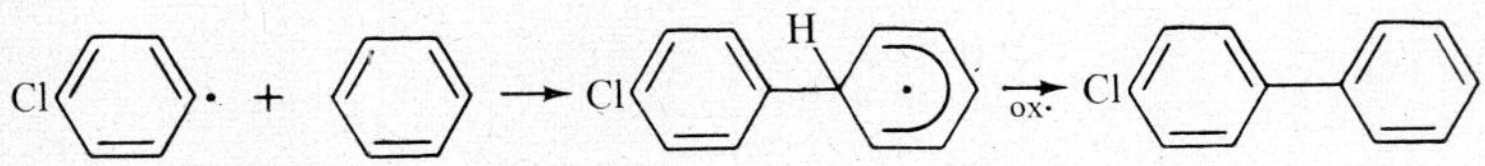

Substituent effects in homolytic aromatic substitution

To describe a substitution of a substituted benzene completely, no matter whether the reaction is radical or ionic, two things must be known: (*i*) the overall rate at which PhX, say, is substituted relative to benzene, which gives the overall activating or deactivating effect of the substituent X, expressed by a rate ratio K^{PhX}_{PhH}. (*ii*) The orientation of the substitution in PhX, *i.e.* the amount of *o*-, *m*- and/or *p*-substitution measured in terms of the isomer ratios α, β, or γ. Combination of these sets of data leads to the derivation of so-called partial rate factors which are a convenient numerical expression of the influence of a given substituent X on the substitution reaction.

Partial rate factors. A partial rate factor f_x is defined as the rate of substitution at position x relative to any position in benzene, this being taken as unity. If the partial rate factors are f_o (for the ortho-position) f_m and f_p, the fraction of *o*-substitution is α, of *m*- is β and of *p*- is γ, and if the reactivity of *each* position in benzene towards the substituting reagent is arbitrarily taken to be unity, then

$$\text{total reactivity of PhX} = 6, \text{ and}$$
$$\text{total reactivity of PhX} = \text{sum of reactivities of each position}$$
$$= 2f_o + 2f_m + f_p$$
$$\text{Hence } K^{PhX}_{PhH} = \frac{2f_o + 2f_m + f_p}{6}$$

The fraction of *o*-substitution α, is given by

$$\alpha = \frac{\text{Reactivity of the } o\text{-positions in PhX}}{\text{Sum of reactivity of all positions in PhX}}$$

Therefore

$$\alpha = \frac{2f_o}{2f_o + 2f_m + f_p} = \frac{2f_o}{6K^{PhX}_{PhH}}$$

$$\text{Hence } f_o = 3\alpha K^{PhX}_{PhH}$$

$$\text{and } f_m = 3\beta K^{PhX}_{PhH}$$

$$\text{and } f_p = 6\gamma K^{PhX}_{PhH}$$

Thus partial rate factors may be obtained from experimentally determined quantities, *viz.* the isomer ratios and rate ratios for substitution in PhX *vs* PhH.

Nowadays such data can be obtained fairly readily as a result of the widespread availability and high accuracy of gas–liquid chromatographs.

Partial rate factors for radical phenylation. Isomer ratios, overall rate ratios and the resultant partial rate factors for a series of mono-substituted aromatic compounds are given in Table 3.

Table 3. Phenylation of aromatic compounds

(At 80°C using 0.2M dibenzoyl peroxide)

X in PhX	% *o*-	*m*-	*p*-	K^{PhX}_{PhH}	Partial rate factors f_o	f_m	f_p
H	—	—	—	1	1	1	1
Cl	50	32	18	1.4	2.2	1.4	1.6
Me	67	19	14	1.7	3.5	1.0	1.4
NO_2	62	10	28	2.9	5.5	0.9	4.9

The striking thing about these results is that, *regardless of the polarity of the substituent group, X,* all substituents have a very small effect on the rate of substitution at each position compared with the rate of substitution of a position in benzene. Further, all positions are substituted to a significant extent. This agrees with the concept of a neutral, uncharged free phenyl radical and is in marked contrast to ionic nitration of these compounds where the very large polar influences of the substituent groups become important (Table 4).

Table 4. Nitration of aromatic compounds

X in PhX	K^{PhX}_{PhH}	f_o	f_m	f_p
H	1	1	1	1
Cl	0.033	0.03	0	0.14
Me	23	40	3	51
NO_2	v. low	0	v. low	0
$CO_2C_2H_5$	0.0037	0.0026	0.0079	0.0009

Those effects which do show up in radical phenylation are clearly second order only. Nitrobenzene displays the largest effect and it has been argued that the higher *o*- and *p*-partial rate factors in this case reflect the possibility of additional delocalization of the single electron into the conjugated nitro group.

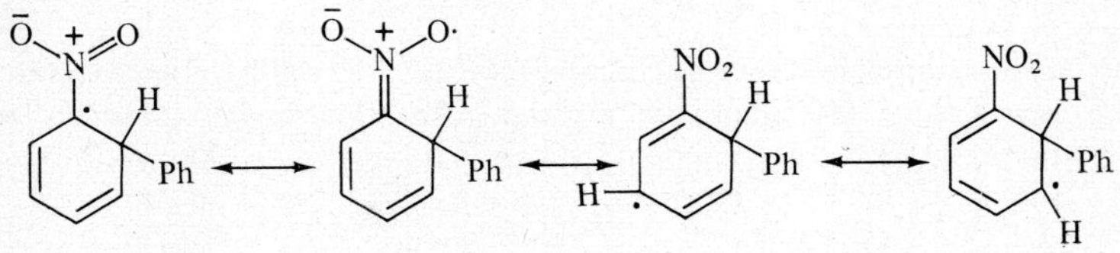

Such additional delocalization is not possible in the case of *m*-substitution.

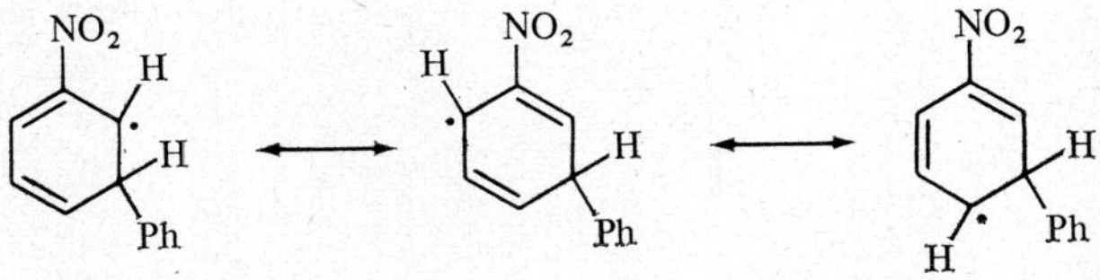

Nevertheless the secondary effects recorded in Table 3 have not been completely explained and it is possible that these small deviations from unity are due to a combination of small steric, polar and delocalizing factors affecting not only the ease of formation and oxidation of the various isomeric intermediate phenylcyclohexadienyl radicals, but also the rates at which these dimerize and disproportionate to give side products. These rates might be expected to differ for the cyclohexadienyl radicals corresponding to *o*-, *m*- or *p*-substitution, thus leading to deviations in the partial rate factors from the figure of unity expected if the radical behaves as a purely neutral, one electron species, not subject to steric factors.

8. Radicals in Synthesis and Biosynthesis

Many examples of the use of radicals in synthesis have already been discussed under other headings and these include the two most important technical applications of radicals, namely radical polymerization and autoxidation. This chapter describes some of the remaining more important reactions of radicals which are useful in synthesis.

Aryl radicals from diazonium salts and related reactions: conversion of anilines into biaryls, aryl halides and arenes

In order to convert an aryldiazonium cation (ArN_2^+) into an aryl radical, ($Ar\cdot$), all that needs to be done, in theory, is to add an electron to the diazonium cation.

$$Ar\overset{+}{N}{\equiv}N: + e \longrightarrow Ar\ddot{N}{=}\ddot{N}\cdot$$

This would lead to the radical $ArN_2^{\cdot}$, which would be expected to be extremely unstable and to decompose immediately to an aryl radical and the stable molecule of nitrogen, formation of the latter being the driving force for the decomposition.

$$ArN{=}N\cdot \longrightarrow Ar\cdot + N_2$$

There are several ways of achieving this, some of which are of historical, as well as practical, interest because they date from the days before the recognition of the existence of short lived free radicals in solution; studies of these reactions played a large part in the development of early free radical theory.

These reactions are of practical interest because they offer convenient routes from the readily available anilines to aryl radicals, and hence products derived from them.

Formation of biaryls

The Gomberg reaction. The reaction of a diazotized aniline with benzene in the presence of aqueous sodium hydroxide to give an unsymmetrical biaryl was discovered by Gomberg in 1923.

$$ArN_2^+Cl^- + HO^- + PhH \longrightarrow ArPh + N_2 + H_2O + Cl^-$$

Hey later showed that the reaction proceeds *via* aryl radicals, and it is interesting that Gomberg, of triphenylmethyl radical fame, never suspected that this arylation involved short lived radicals.

The reaction is of variable usefulness, often giving rise to tarry products, and has now been superseded as a route to biaryls by the diaroyl peroxide route (Chapter 4) and by *in situ* aprotic diazotization of anilines (*see* below).

The mechanism of the reaction, elucidated only recently, involves the formation of the aryldiazotate ion. Under the alkaline conditions of the reaction

$$ArN_2^+\bar{O}H \rightleftharpoons ArN{=}NOH \rightleftharpoons ArN{=}NO^- + H^+$$

this diazotate anion is then believed to be the source of the electron needed to reduce ArN_2^+ to $ArN_2^{\cdot}$ and hence Ar·. This occurs either by direct electron transfer

$$ArN_2O^- + ArN_2^+ \longrightarrow ArN_2O\cdot + Ar\cdot + N_2$$

or *via* the intermediate formation and subsequent homolysis of a bis-diazo ether

$$ArN_2O^- + ArN_2^+ \longrightarrow ArN{=}NON{=}NAr \longrightarrow Ar\cdot + N_2 + ArN_2O\cdot$$

(electron transfer)

Arylation of benzene then occurs to give an intermediate aryl-cyclohexadienyl radical which can be oxidized either by $ArN_2O\cdot$ or, better, by unchanged ArN_2^+. In either case an aryl radical is regenerated in a chain process.

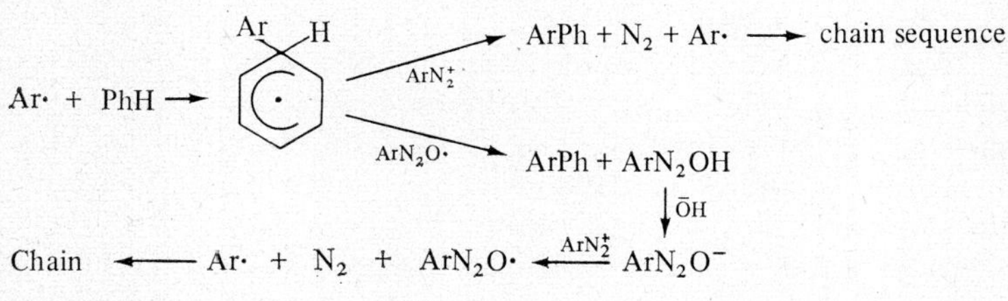

Aprotic in situ *diazotization of anilines.* This much more convenient and efficient modification of the Gomberg reaction involves the *in situ* diazotization of the parent aniline by pentyl nitrite in benzene or another organic solution, rather than in water. Good, one-pot conversions of the aniline to the biaryl (up to 75 per cent) can be achieved in this way.

$$ArNH_2 + RONO \longrightarrow ArN_2^+OR^- + H_2O$$
$$ArN_2^+OR^- + PhH \longrightarrow ArPh + N_2 + ROH$$

This reaction also proceeds *via* aryl radicals, probably by a mechanism similar to that of the Gomberg reaction.

Decomposition of acylarylnitrosamines. These compounds, exemplified by *N*-nitrosoacetanilide, are easily prepared as low melting, unstable, solids by reaction of the parent anilide with nitrous fumes or nitrosyl chloride. More recently they have been prepared more conveniently and with less hazard by reaction *in situ* of pentyl nitrite or *p*-chlorobenzoyl nitrite (p-ClC_6H_4COONO) with the anilide.

$$PhNHCOMe \xrightarrow{NOX} PhN(NO)COMe + HX$$
$$[X = Cl,\ p\text{-}ClC_6H_4CO_2 \text{ or } RO]$$

N-nitrosoacetanilide, on dissolution in a solvent such as benzene rapidly undergoes rearrangement at room temperature to the isomeric benzenediazonium acetate which, in turn, is converted into a phenyl radical by a reaction sequence closely related to that of the Gomberg reaction, the first steps being

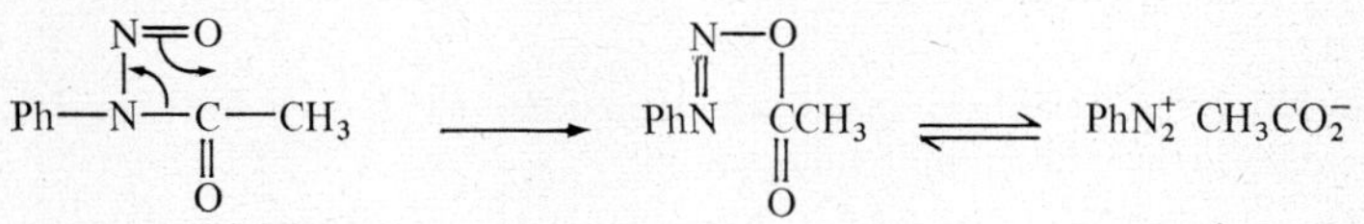

$$MeCO_2^- + PhN_2OCOMe \longrightarrow (MeCO)_2O + PhN_2O^-$$

$$PhN_2^+ + PhN_2O^- \longrightarrow PhN_2O\cdot + N_2 + Ph\cdot \quad \textit{et seq.}$$

In each of three reactions described above the use of a substituted aniline in benzene leads to the corresponding substituted biaryl in which the positional integrity of the substituent is retained, *e.g.* (Chapter 7).

$$p\text{-}ClC_6H_4NH_2 + PhH \xrightarrow{\text{Gomberg}} p\text{-}ClC_6H_4Ph$$

The use of a substituted benzene, however, leads to the formation of all three isomeric biaryls, *e.g.*

$$PhNH_2 + PhCl \xrightarrow{\text{Gomberg}} o\text{-}, m\text{-} + p\text{-}ClC_6H_4Ph$$

Copper(0) *catalyzed internuclear arylation: the Pschorr reaction.* The single electron needed to convert ArN_2^+ into Ar· can be provided by suitable multivalent metals such as Cu^0, Cu^+, Fe^{2+}, Zn^0. This has been used in the Pschorr reaction of bicyclic diazonium salts of suitable geometry to lead to the formation of new rings *via* internuclear cyclization, *e.g.* the conversion of 2-diazo-α-phenyl-*cis*-cinnamic acid into phenanthrene-9-carboxylic acid. There are many heterocyclic extensions of this reaction.

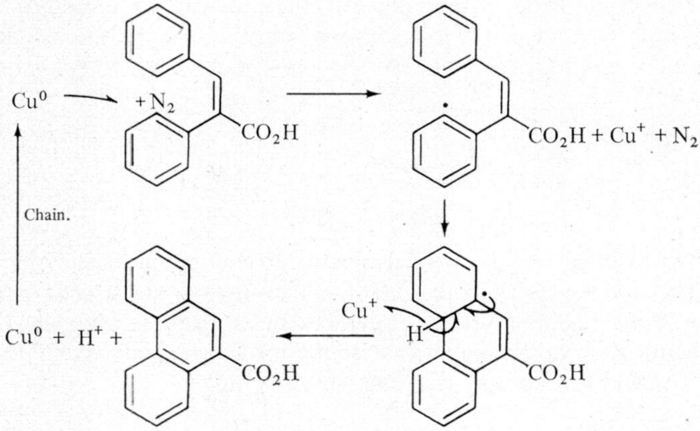

This reaction is a chain reaction because Cu^0 is returned to the system at the end of each cycle.

Arylation of olefinic double bonds: the Meerwein reaction. Reaction of aryldiazonium salts with olefins such as ethyl cinnamate in the presence of a mixture of cuprous and cupric ions proceeds first *via* reduction of the diazonium cation to the aryl radical by cuprous ions, by analogy with the Pschorr reaction. The aryl radical then adds to the olefinic double bond and the resulting intermediate radical is oxidized by cupric ion.

$$ArN_2^+ + Cu^+ \longrightarrow Ar\cdot + N_2 + Cu^{2+}$$

$$Ar\cdot + PhCH{=}CHCO_2Et \longrightarrow Ph\dot{C}HCH\begin{cases}Co_2Et\\Ar\end{cases}$$

$$\downarrow Cu^{2+}$$

$$PhCH{=}C\begin{cases}Co_2Et\\Ar\end{cases} \xleftarrow{-H^+} Ph\overset{+}{C}HCH\begin{cases}Co_2Et\\Ar\end{cases} + Cu^+$$

Formation of aryl halides

The Sandmeyer and Gatterman reactions. The long known Sandmeyer conversion of arenediazonium halides into the corresponding aryl halides, under the catalytic influence of cuprous halides also proceeds *via* a series of single electron transfer redox reactions. On allowing the two reactants, *e.g.* aqueous benzene diazonium chloride and cuprous chloride, to mix a precipitate of a diazonium chloride–cuprous chloride complex is formed. Subsequent reaction proceeds within this solid complex with gradual evolution of nitrogen and the formation of chlorobenzene. The reaction proceeds in a cyclic fashion by a series of redox steps as indicated below (semi headed arrows indicating the movement of one electron).

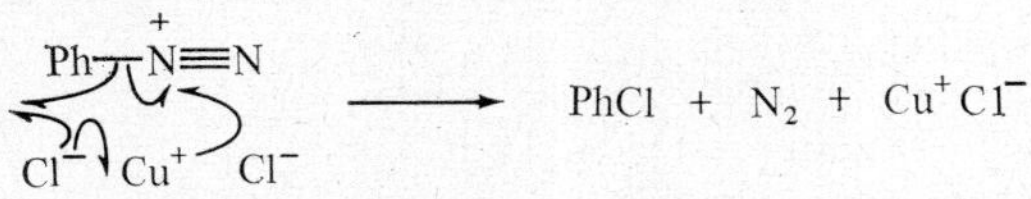

As to be expected from this mechanism, cuprous bromide and the corresponding diazonium bromide are required in order to obtain aryl bromides: 'mixed' halogens such as cuprous bromide with a diazonium chloride lead to a mixture of aryl bromide and chloride.

The Gattermann procedure is related to the Sandmeyer reaction. It is now little used because the yields are variable. Here, copper(0) is the source of the needed electron.

$$ArN_2^+ + Cu^0 \longrightarrow Ar\cdot + N_2 + Cu^+$$
$$Ar\cdot + Cu^+ + Cl^- \longrightarrow ArCl + Cu^0$$

Formation of aryl bromides by aprotic diazotization of anilines in bromoform. This useful route to aryl bromides uses the aprotic *in situ* diazotization of anilines by pentyl nitrite, described above, to give aryl radicals. These then abstract a bromine atom from bromoform to give the aryl bromide.

$$Ar\cdot + CHBr_3 \longrightarrow ArBr + \cdot CHBr_2$$

In principle, the reaction is capable of extension to the synthesis of aryl chlorides, but separation difficulties lead to lower yields than are desirable.

Conversion of anilines into arenes

Reduction of the diazonium group. The most widely used method involves reaction of the aqueous diazonium salt with an excess of hypophosphorous acid. The reaction involves an aryl radical, although the nature of the initiation step is obscure, which abstracts hydrogen from hypophosphorous acid to give the reduced product and the hypophosphite radical. This, in turn, is oxidized by unreacted diazonium cation to give hypophosphite positive ion and a new aryl radical, which continues the chain.

$$Ar\cdot + HP(OH)_2 \longrightarrow ArH + \cdot P(OH)_2$$
$$\cdot P(OH)_2 + ArN_2^+ \longrightarrow Ar\cdot + N_2 + {}^+P(OH)_2$$
$${}^+P(OH)_2 + H_2O \longrightarrow P(OH)_3 + H^+$$

Recently, a variant of the aprotic *in situ* diazotization of anilines by alkyl nitrites has been used successfully in a similar way. In the presence of tetrahydrofuran, which contains easily abstracted α-hydrogen atoms, the following chain reaction occurs.

$$ArNH_2 + RONO \longrightarrow ArN_2^+OR^- + H_2O$$

$$ArN_2^+OR^- \longrightarrow Ar\cdot$$

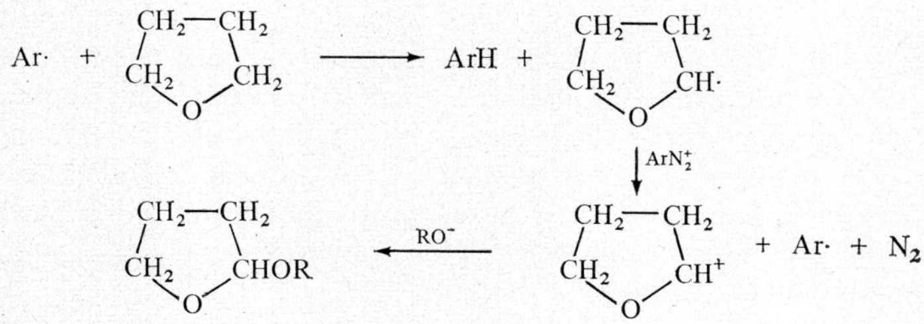

Photolysis of alkyl nitrites: the Barton reaction

Photolysis of nitrites occurs readily under the influence of relatively low energy radiation ($\lambda = 300$–350 nm) to give alkoxyl radicals and nitric oxide.

$$RONO \longrightarrow RO\cdot + NO$$

The alkoxyl radical, regardless of its mode of generation, has a high reactivity towards aliphatic and alicyclic hydrogen atoms, present in excess.

$$RO\cdot' + R'H \longrightarrow ROH + R\cdot'$$

Barton showed that in the photolytic reaction in suitably constructed molecules, *intramolecular* hydrogen abstraction is a highly favoured reaction if there is a possibility of a six membered transition state.

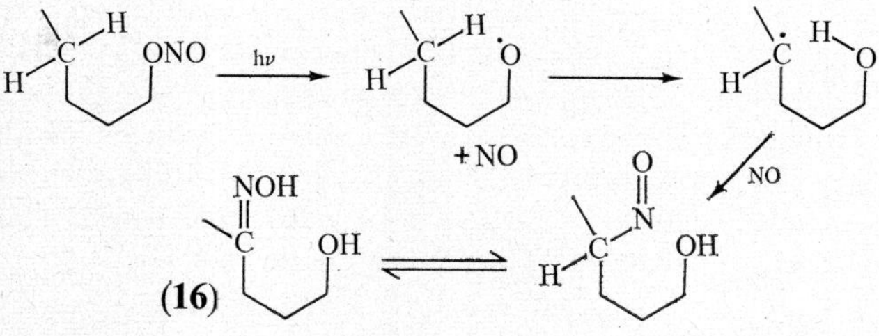

(15) **(16)**

The intermediate alkyl radical (**15**), produced by intramolecular hydrogen abstraction, reacts with the nitric oxide formed in the photolysis to give the corresponding nitrosoalkane which tautomerizes to give the oxime (**16**), which can then be converted, for example, into the corresponding carbonyl compound. The reaction is not a chain process and requires photolysis of one molecule of starting material to produce one molecule of product, *i.e.* a quantum efficiency of unity. It is, in fact, a photoisomerization involving the intermediacy of alkoxyl radicals. The importance of the sterically favoured six-membered intermediate is emphasized by the results of photolysis of long chain alkyl nitrites which, despite offering several possible points of intramolecular attack, give exclusive reaction *via* a six-membered intermediate, *e.g.*

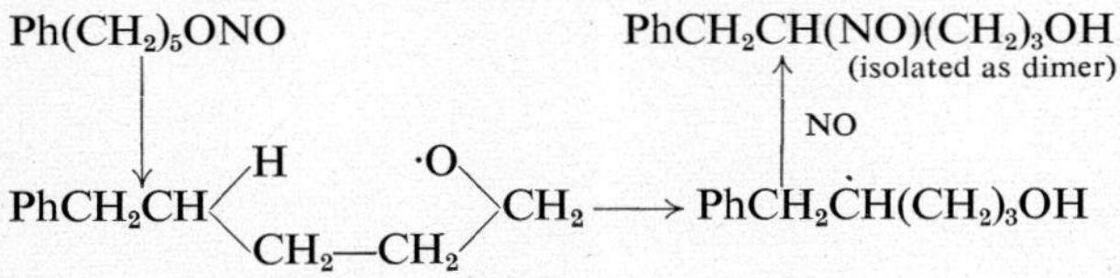

In effect, this simple reaction transfers functionality from position 1 to an unactivated position 5 in the molecule. Barton used this to functionalize carbon 18 in the steroid series using readily available starting materials. This is well illustrated by the three step synthesis of the hormone aldosterone (**17**), hitherto very difficult to synthesize by conventional methods. The synthesis involves the conversion of the reasonably accessible alcohol (**18**) into the nitrite ester and hence *via* the oxime (**19**) into aldosterone acetate (**17**). Thus Barton was able to make 40 or 50 g of aldosterone at a time when the previous total world supply was only a few mg.

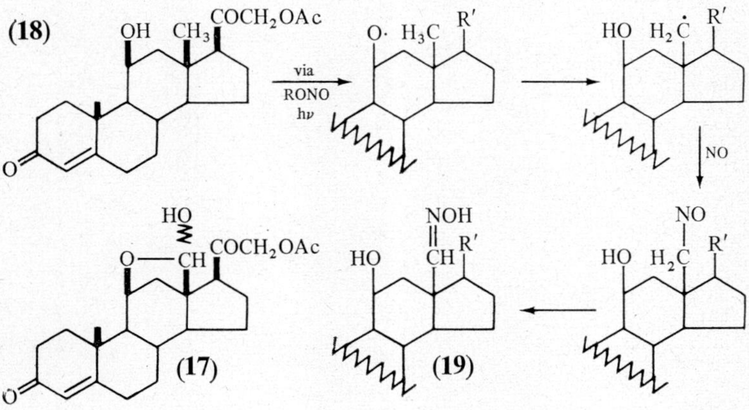

(R′ = COCH2OAc)

The Barton reaction has been widely used in the synthesis of steroids, many of which are medically important. Other routes to steroidal alkoxyl radicals, such as oxidation of the parent alcohol by lead tetraacetate, or photolysis of hypochlorites (R′OCl → R′O· + Cl·) have also been used.

Oxidation of phenols: biogenesis of phenolic alkaloids

As stated earlier (Chapter 3), one-electron oxidation of phenols gives aryloxyl radicals, which, by analogy with other benzylic type radicals, would be expected to be radicals of intermediate lifetime by virtue of the delocalization of the single electron which is possible.

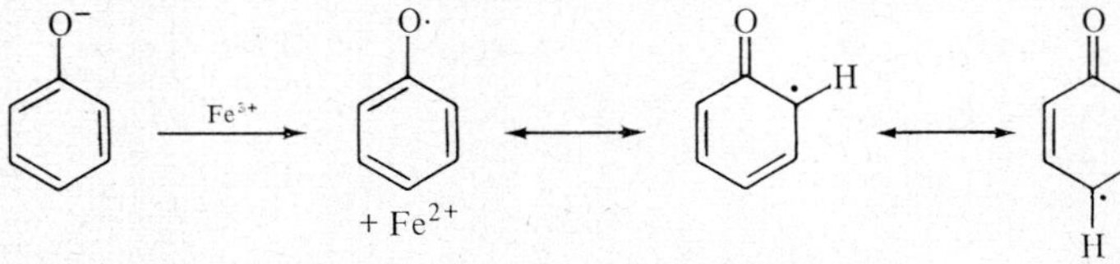

Such radicals react by C–C coupling which may involve *otho-ortho*, *ortho-para*, or *para-para* reaction. O-O Coupling, which would give the unstable peroxide, is not observed.

Although oxidation of phenols has been known for a long time it is only in comparatively recent times that the mechanism of the reaction, its scope in synthesis, and its significance in biogenesis have been elucidated. Thus Pummerer in 1925 oxidized *p*-cresol and obtained a crystalline dimer, 'Pummerer's ketone', the incorrect structure of which remained unchallenged until Barton proved that the structure was (**20**), and put forward a mechanism based on *o-p* coupling of *p*-cresyloxyl radicals.

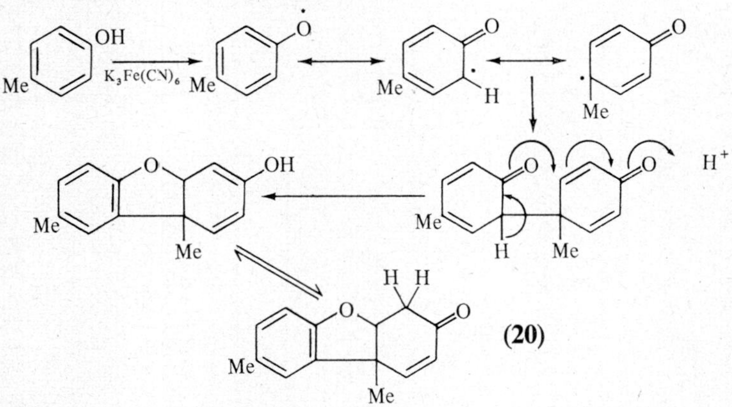

(**20**)

Barton then used this concept to devize a two-step synthesis of the plant lichen product usnic acid (**21**).

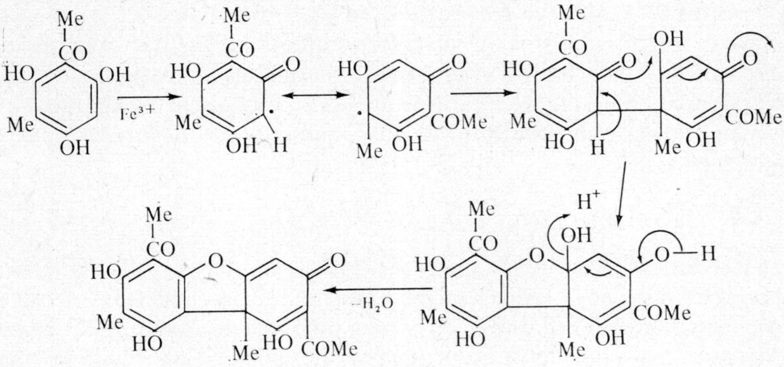

(21)

This synthesis lent weight to the idea previously put forward by Barton, Battersby, Erdtman and others that biogenesis of certain plant alkaloids involved similar coupling of aryloxyl radicals. For example, it was suggested in 1925 by Robinson that the morphine alkaloids, which had been of interest since the beginning of the last century, were formed in nature by oxidative cyclization of a benzylisoquinoline (**22**). The establishment of the details of the mechanism had to wait another 20 years, but it is now certain that the benzylisoquinoline (**22**) undergoes oxidation followed by *intra*molecular *o-p* coupling (*cf* the case of Pummerer's ketone) to give the dienone (**23**). Intramolecular condensation then gives the enone (**24**) which on reduction can lead eventually to codeine (**25**; R′ = Me) or morphine (**25**; R′ = H).

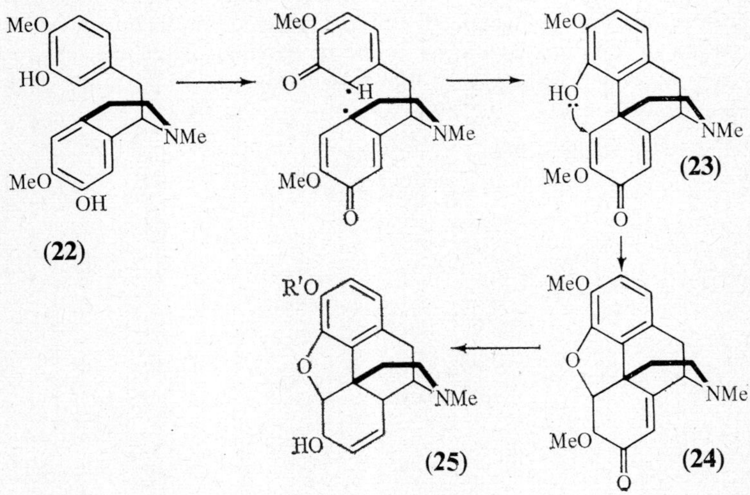

(22) (23) (24) (25)

Electrolytic reactions of organic compounds

This is a growing area of organic synthesis. The first example is provided by the Kolbe electrolysis of aqueous potassium acetate, discovered in 1849, wherein acetate ions discharged oxidatively at the anode are converted into acetoxyl radicals and hence *via* methyl radicals to ethane.

$$2MeCO_2^- \xrightarrow{-1e} 2MeCO_2^{\cdot} \longrightarrow 2Me\cdot \longrightarrow C_2H_6$$

In the intervening years many examples of electrolytic reduction and oxidation have been described and, since these reactions involve the formal transfer of one electron at a time, the intermediates formed in these reactions have radical characteristics. For example, the industrially important electrohydrodimerization of activated alkenes proceeds *via* one electron reduction to give a radical anion, which can either dimerize or react with another molecule of alkene in a Michael type reaction.

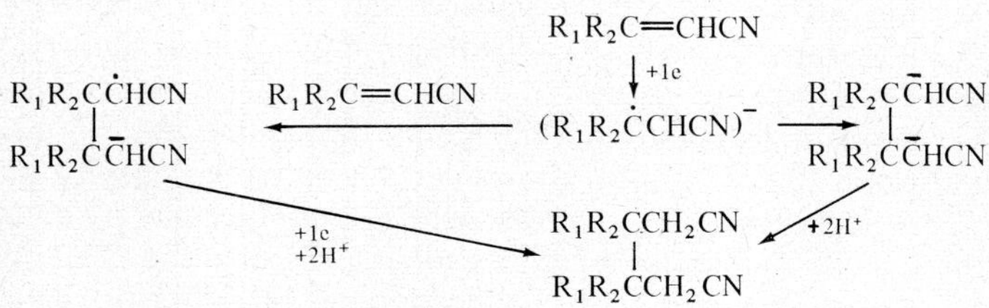

The product is a derivative of adiponitrile, a nylon precursor.

In general it is possible to explain most electrolytic oxidations and reductions in terms of radical intermediates, but often the picture is complicated by the heterogeneous nature of the electrode reaction. There is no doubt, however, about the importance of this fast growing field.

9. Carbenes and Nitrenes

The chemistry of carbenes and nitrenes is very wide indeed and a full account is outside the scope of this book. Nevertheless, since certain reactions of these reactive intermediates are radical-like a very short description is included in this Chapter to serve as a basis for further reading.

Carbenes

Carbenes are neutral divalent carbon species, carrying a sextet of electrons. The carbon atom is covalently bonded to two other groups, thus leaving two non bonded electrons to be considered, *e.g.* $:CH_2$ or $:CCl_2$. These two electrons may be spin paired (↑↓), with opposing spins, in which case the carbene is in the singlet state, or they may have parallel spins (↑↑), in which case the carbene is in the triplet state. These terms arise as follows: the spin value of an electron is $\frac{1}{2}$ and if the total spin is S, the spin multiplicity is given by $2S + 1$. Thus for spin paired electrons (↑↓) $S = +\frac{1}{2} - \frac{1}{2} = 0$, hence the spin multiplicity is 1 and the species is singlet. For parallel spins (↑↑), $S = +\frac{1}{2} + \frac{1}{2}$, therefore $(2S + 1) = 3$, hence the term triplet.

There are several possible structures for carbenes depending on (*a*) the spin multiplicity and (*b*) whether the carbene, be it singlet or triplet, is in the ground state or an excited state. Dichlorocarbene, $:CCl_2$, produced by action of base on chloroform

$$CHCl_3 \xrightarrow{HO^-} {}^-CCl_3 \longrightarrow Cl^- + :CCl_2$$

is considered to be in the lowest singlet state having an sp^2 configuration in which the spin-paired electrons occupy one lobe of the sp^2 orbital leaving a vacant p_z orbital. The species therefore behaves as an electrophile, reacting, for example, by cycloaddition in a

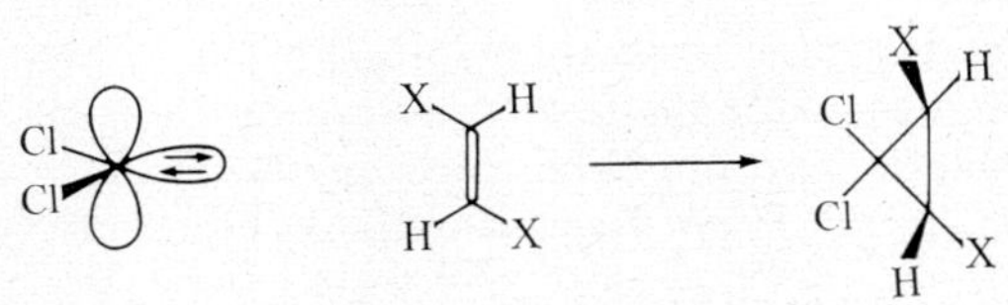

stereospecific manner with *cis*- or *trans*-alkenes to give cyclopropanes.

Methylene, $:CH_2$, on the other hand, produced by photolysis of diazomethane in the presence of benzophenone, as a photosensitiser, has triplet multiplicity and in the ground state can also be linear, with the unpaired electrons occupying orthogonal *p* orbitals.

Regardless of the shape of triplet methylene, however, the important point concerning its chemical behaviour lies in the unpaired nature of the free electrons. The fact that these are not coupled leads to diradical behaviour in which each single electron has radical character. Thus reaction of triplet methylene with an alkene gives a cyclopropane in a non stereospecific reaction, indicating stepwise, rather than concerted, reaction of the diradical. Radical abstraction reactions also occur.

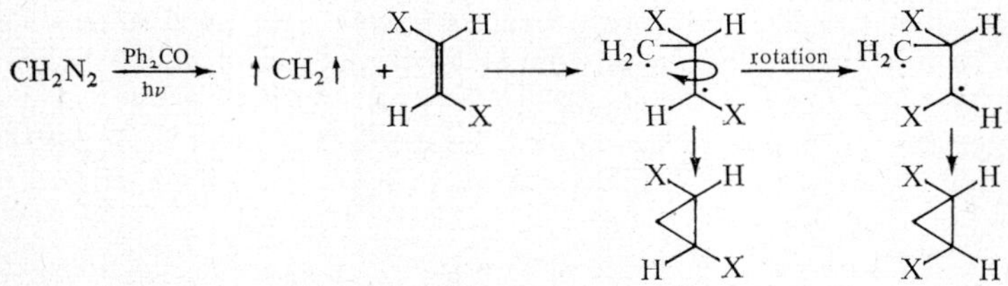

Unless steps are specifically taken to produce triplet carbenes, as in the preceding case, they are generally produced in an excited singlet form in which they may react. Alternatively, depending on molecular circumstances, they may degenerate by collision to a triplet ground state.

Nitrenes

These are neutral monovalent nitrogen species and are the direct isoelectric analogues of carbenes. Like carbenes these may also exist in singlet (↑↓) and triplet (↑↑) forms, depending on molecular circumstances and experimental conditions. The former are electrophilic in character while the latter undergo radical-type abstraction reactions. Nitrenes are usually formed by photolysis or thermolysis of azides or by deoxygenation of nitro compounds by tervalent phosphorus reagents.

$$ArN_3 \longrightarrow ArN\colon + N_2$$

$$ROOCN_3 \longrightarrow RO_2CN\colon + N_2$$

$$ArNO_2 + 2(RO)_3P \longrightarrow ArN\colon + 2(RO)_3PO$$

Like the corresponding carbenes, singlet nitrenes such as RO_2CN (↑↓) tend to add stereospecifically to double bonds whereas in the

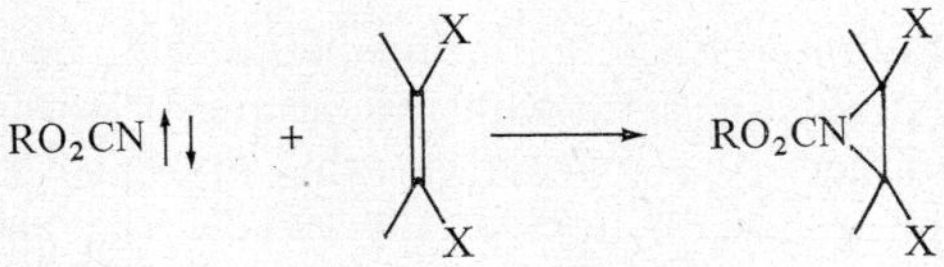

triplet form the addition is non stereospecific and is in competition with radical abstraction reactions involving neighbouring C–H bonds.

10. Detection of Radicals

The participation of radicals in reactions may be inferred from correlation of reaction rates with the nature of substituents and/or from the products of the reaction—this approach has been very successful and underlies most of the radical chemistry which has been described in previous chapters. Such evidence has been complemented by the physical detection of radicals in some cases. The best physical methods are based on two types of magnetic resonance spectroscopy: (*a*) chemically induced dyanamic nuclear polarization spectroscopy (CIDNP), sometimes known as 'kidnap' spectroscopy, which is carried out in the cavity of a nuclear magnetic resonance spectrometer, and (*b*) electron spin resonance spectroscopy (esr), sometimes known as electron paramagnetic resonance spectroscopy (epr). Although these forms of absorption spectroscopy are applicable in completely different situations they have fundamental features in common, the most important aspect being that electrons and the nuclei of certain atoms are said to spin. Since they also carry charge this leads to the generation of a magnetic moment. In other words, these species behave like tiny bar magnets which will respond to a magnetic field. Magnetic resonance spectroscopy is based on these magnetic properties of the electron and certain nuclei.

Nuclear magnetic resonance (nmr) spectroscopy

Nuclei having magnetic moments will take up various alignments in a magnetic field, the alignment of lowest energy is clearly that in which the nucleus is exactly lined up with the field; other alignments are less stable (a good analogy is that the position of the compass needle is most stable when pointing north). On the molecular scale, however, quantum theory allows only certain alignments corresponding to certain levels of energy. The number of possible orientations is given by a spin quantum number designated S for the electron ($S = \frac{1}{2}$), and I for a nucleus [$I = 0, \frac{1}{2}, 1, 1\frac{1}{2} \ldots$]. Nuclei such as ^{12}C, ^{16}O do not have a magnetic moment and $I = 0$. These do not give a magnetic resonance spectrum. Nuclei such as ^{1}H, ^{19}F and ^{31}P have $I = \frac{1}{2}$ and can take up two orientations only in a magnetic field, *i.e. with* the field (low energy state α) and *against* the field (high energy state β).

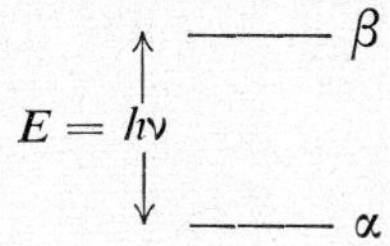

In practice a proton in a magnetic field of 23 kgauss takes up *both* orientations but there are very slightly more in the low energy state α, the difference being only 1 in 10^6. If the proton is now irradiated with radiation of suitable energy (frequency ν; $E = h\nu$, in this case ν = 100 MHz, corresponding to $E = 0.04$ J mol^{-1}) it is possible to change from the low energy alignment to that of high energy and *vice versa*, *i.e.* to 'flip' the magnet over. The energy required is quantized and corresponds to the difference between the potential energies of the two levels. This absorption of energy leads to the observed nuclear magnetic resonance spectrum. It is an *absorption* (α → β) rather than an *emission* (β → α) because slightly more protons are in the low- rather than the high-energy state in the magnetic field. The frequency of irradiation (and hence absorption) depends on the field and is given by $\nu = \gamma \boldsymbol{H}_0/2\pi$ where $\boldsymbol{H}_0$ is the field in gauss and γ is a nuclear constant: the gyromagnetic ratio ($\gamma = 26\,750$ for ^{1}H). For protons in a field of 23 490 gauss the energy needed to flip over the proton is in the radiofrequency region and corresponds to a frequency of 100 MHz. In practice, protons in different environments have slightly different resonance frequencies giving rise to different signals in the spectrometer. From

FIG. 3. A typical ^{1}H nmr spectrum.

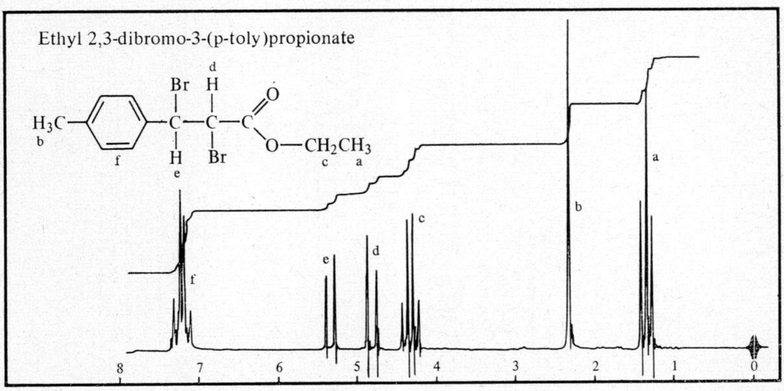

the number and splitting of the signals the number of different types of protons in a molecule can be determined, the positions of the signals indicate the electronic environment, while the intensities of the signals show how many of each type of proton are present. Since similar considerations apply to other nuclei with $I = \frac{1}{2}$, nuclear magnetic spectroscopy is a very powerful tool for determination of the structures of molecules.

Electron spin or paramagnetic resonance (esr or epr) spectroscopy

The odd electron in a free radical has a spin ($S = \frac{1}{2}$) and hence it has a magnetic moment. By analogy with the proton, the metaphorical radical-magnet can align with or against an external magnetic field, and energy, in the form of irradiation of the correct frequency, will change the alignment from the one energy state to the other.

Thus an esr spectrum can be produced which is analogous to a nmr spectrum produced from nuclei such as ^{1}H, ^{19}F, ^{31}P *etc.* The electron, however, has a much larger magnetic moment than a proton, say, hence more energy is required to reverse the spin. Thus, whereas spin transitions in pmr at fields of 14 000 gauss can be effected in the radiofrequency region (60 MHz), the corresponding esr signals in a field of 13 000 gauss require microwave radiation

FIG. 4. A typical esr spectrum.

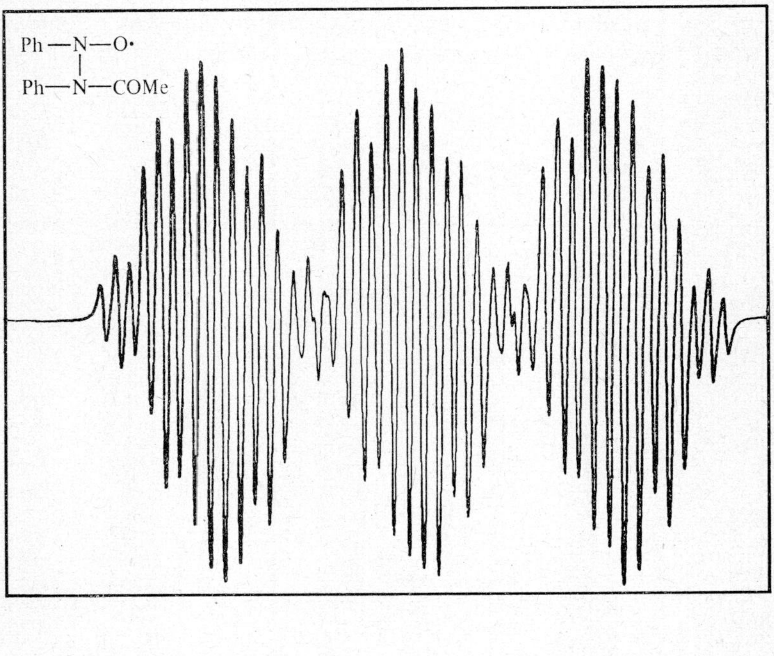

(36 000 MHz). In practice a field of about 3000 gauss (~9000 MHz) is used. Esr and nmr signals are therefore produced in similar fashions by irradiation of a sample in a magnetic field with the difference that esr signals are usually presented, for reasons of sensitivity, as the first derivative of the signals. Therefore, provided the radical is present in a sufficiently high concentration, an esr signal can be obtained, so providing direct physical evidence of the presence of a radical. Further, the nature and position of the signal enables the structure of the radical to be determined.

Absorption of energy will occur when $h\nu = g\beta\boldsymbol{H}$, where ν is the frequency of irradiation, β is the Bohr magneton, $\boldsymbol{H}$ is the field strength and g the spectroscopic splitting factor, also known as the g value, which measures the contribution of spin and/or total angular momenta to the total angular momentum. For a free electron $g = 2.00232$ but for most radicals g varies only slightly (± 0.01) from this value according to the structure of the radical.

The method is very sensitive and can detect radicals at concentrations as low as 10^{-10}M. Clearly the longer the life of the radical, the better the chance of observing its spectrum. Long lived species such as triphenylmethyl, various nitroxides, Galvinoxyl, *etc.* are easily observed in static systems but more reactive radicals such as Ph·, alkyl, benzyl are more difficult to observe. In these cases one may use flow systems whereby the radicals are generated in the spectrometer cavity in a continuous fashion thus maintaining a stationary concentration, or the radicals may be produced by photolysis of the radical precursor in a frozen matrix, which prevents collision and hence removal of the radicals by reaction. A third alternative is to produce a sufficiently high stationary concentration of radical by continuous irradiation of the radical precursor with fast electrons (2.8 MeV). This technique enables the esr spectrum of reactive radicals such as alkyl to be recorded in the liquid phase, and variations of this method of production of the necessary stationary concentration of radicals based on continuous photolysis and thermolysis are now becoming increasingly used.

Another device is the use of *spin traps* whereby short lived radicals, *e.g.* Ph·, difficult to observe by normal esr means, are trapped by nitroso compounds to give nitroxide radicals of long life which can be observed at leisure in the esr spectrometer.

$$\mathrm{Ph\cdot} + \mathrm{Me_3CN{=}O} \longrightarrow \begin{matrix}\mathrm{Ph}\diagdown \\ \mathrm{Me_3C}\diagup\end{matrix}\mathrm{N{-}O\cdot}$$

Not only can the spin trapping technique be used to show the presence of radicals in the system, but it is also possible by analysis of the signal produced in many cases to determine the structure of the primary radicals involved.

The ease with which nitroxide radicals can be detected by esr spectroscopy has led to the development, by McConnell's school, of the technique known as *spin labelling* to monitor molecular changes which occur in biological systems. This consists of allowing a nitroxide such as (**26**) to react *via* a non radical reaction of its constituent isocyanate group with the amino or hydroxyl groups of a protein, say, the protein nitroxide is then capable of giving an esr signal, which can be observed during biological changes,

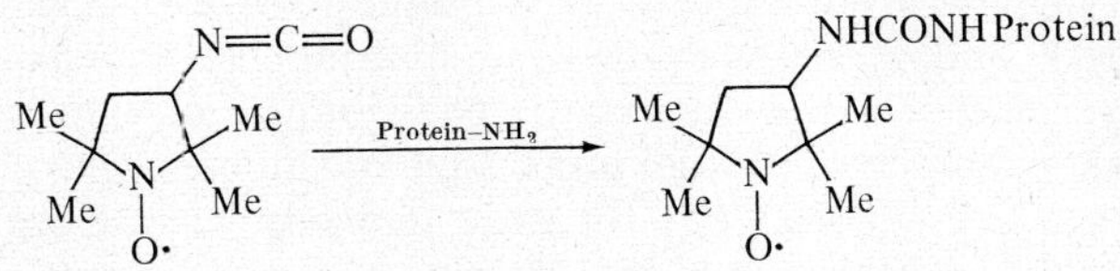

(**26**)

thus giving valuable information about protein behaviour. In the case of bovine serum albumin the changes in the nitroxide signal with change in pH have been related to the actual uncoiling of the protein during the reaction.

Chemically induced dynamic nuclear polarization spectroscopy (CIDNP)

This phenomenon, discovered in the late sixties, occurs during nmr investigation of molecules produced from radical reactions, in certain but not all circumstances. *I.e.* CIDNP is observed in the *products* of certain radical reactions rather than in the radicals themselves (as in esr). Obviously, it has great diagnostic potential in radical chemistry.

It has already been shown that in normal absorption pmr spectroscopy the relative intensities of the signals (as measured by the areas of the peaks) are proportional to the relative concentrations of the corresponding types of protons. When CIDNP occurs in radical-derived products it manifests itself in either of two ways: (*a*) by a much *enhanced absorption*, *i.e.* it gives rise to an absorption peak much greater than that expected on the basis of the number of protons present, and (*b*) by *emission*, which is also enhanced on a proton count basis, as shown by an inverted peak on the spectrogram. Depending on the molecular circumstances and experimental conditions, CIDNP may not appear at all in some radical reactions but when it does it may appear as enhanced absorption or as enhanced emission. In any event it decays to the 'normal' absorption spectrum in due course (about 10 min is considered to be a relatively long time).

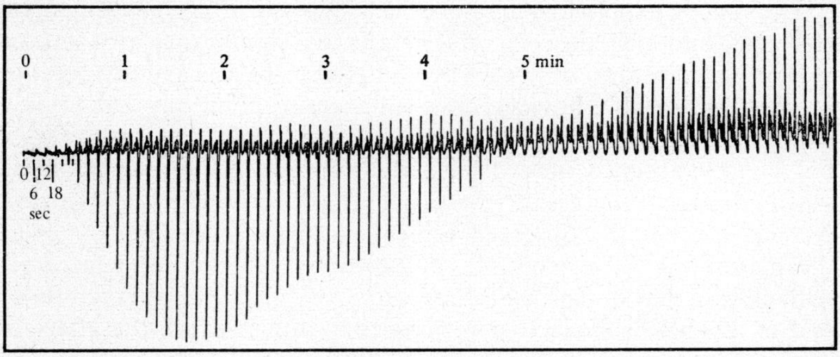

FIG. 5. The CIDNP spectrum of PhH produced from benzoyl peroxide in cyclohexanone at 100 °C. Signals recorded at 6 sec intervals in the region of 2.6 °C.

The appearance of this phenomenon in radical derived products is related to the very nature of radical reactions. As shown throughout this book, radical processes involve the pairing of single electrons; single electron paramagnetic species have a spin, and hence a magnetic moment, therefore there will be magnetic resonance consequences associated with this. Also the 'normal' appearance of magnetic resonance signals arising from a proton in a molecule is a function of the relative population of the two possible spin states (α and β) of the proton involved, and in 'normal' cases, $\alpha > \beta$, hence absorption is observed. If the product is derived from a radical reaction, however, there is now the possibility that the normal relative occupancy of the two spin states will be disturbed because the radical precursor of the product possessed electron spin, and hence magnetic moment. For example, the β level may be more highly occupied than the lower α level, thus leading to emission.

A good example of CIDNP is provided by the decomposition of dibenzoyl peroxide in cyclohexanone, in which the signal due to the aromatic protons in benzene produced by abstraction of a

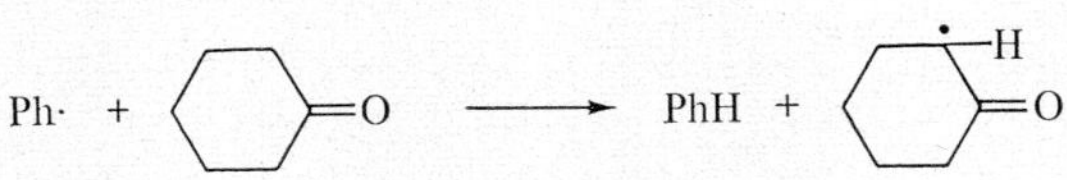

hydrogen atom from cyclohexanone, appears first as an enhanced emission decaying to zero and gradually with time reappearing as an increasing absorption to the 'normal' signal, the entire process taking up to 10 min according to conditions.

The theory explaining the possible interactions between unpaired electrons and neighbouring nuclei and its significance in relation to the observations of CIDNP is complex.* In the context of the present monograph, however, it is sufficient to point out that (*a*) not all radical reactions exhibit CIDNP, *i.e.* the absence of CIDNP does not signify absence of radicals, and (*b*) that the observation of CIDNP does not necessarily imply that the product is the major product or that the observed radical pathway is the only or even the major one. The strengths of CIDNP as a technique compared with esr are that it can be more sensitive, by an order of magnitude, and is much less subject to misinterpretation because the effect is observed in the product. Further, it is not so dependent on the lifetime of the radical.

Acknowledgments

It is a great pleasure to record my gratitude to a number of colleagues and friends: to my teacher, D. H. Hey, FRS, who first aroused my interest in this subject and sustained it with his advice and help when things went wrong, and to Dr J. H. Knox, Dr A. Ledwith, Dr I. H. Sadler and Professor C. J. M. Stirling, who were kind enough to read the entire manuscript and who made many helpful comments.

J. I. G. Cadogan

May, 1973

---

* The time of preparation of this monograph was one of intense activity and development of the field, the most significant advance being the formulation of Kaptein's rules governing the phenomenon.

Appendix: Bond Dissociation Energies

in kJ mol^{-1} (kcal mol^{-1})

Me–H	422 (101)	$BrCH_2$–H	413 (99)
C_2H_5–H	410 (98)	Cl_3C–H	376 (90)
		OCH–H	317 (76)
Me_2CH–H	392 (94)	Ph–H	426 (102)
Me_3C–H	371 (89)	$PhCH_2$–H	356 (85)
		$\cdot CH_2CH_2$–H	167 (40)
Me–Me	346 (83)	Me–$\dot{C}O$	71 (17)
Me–CHO	355 (85)	$PhCH_2$–CH_2Ph	196 (47)
$CH_2{=}CH_2$	522 (125)		
Me–F	446 (107)	Cl_3C–Cl	284 (68)
Me–Cl	334 (80)	Cl_3C–Br	204 (49)
Me–Br	280 (67)	CH_3–OH	376 (90)
Me–I	222 (53)	$PhCH_2$–Br	211 (51)
MeS–Me	304 (73)	MeN=NMe	192 (46)
MeHg–Me	213 (51)		
H–H	430 (103)	H–F	560 (134)
H–OH	501 (120)	H–Cl	430 (103)
H–OOH	376 (90)	H–Br	363 (87)
HO–OH	212 (51)	H–I	296 (71)
H–NH_2	426 (102)	F–F	150 (36)
H_2N–NH_2	251 (60)	Cl–Cl	238 (57)
MeO–NO	150 (36)	Br–Br	192 (46)
MeO–OMe	155 (37)	I–I	150 (36)
Me_3CO–$OCMe_3$	155 (37)	N≡N	940 (225)
MeCOO–OCOMe	125 (30)	MeS–SMe	305 (73)

Suggestions for Further Reading

General

D. H. Hey and W. A. Water's now classic review (*Chem. Rev.*, 1937, **21**, 169) is well worth study. Very good complementary texts are available; in particular Cheves Walling's *Free radicals in solution* (London: Wiley, 1957) is a splendid source book. Also recommended are C. J. M. Stirling's *Radicals in organic chemistry* (London: Oldbourne, 1965), W. Pryor's *Free radicals* (New York: McGraw Hill, 1966) and E. S. Huyser's *Free radical chain reactions* (New York: Wiley, 1970). W. Pryor has also written an excellent concise article on organic free radicals (*Chem. Engng. News*, 1968, 70)

Up to date specialist articles on the latest advances and thinking in radical chemistry are available in two collections: *Essays on free radical chemistry* (*Chem. Soc. Spec. Pub.*, 1970, No. 24) issued to mark the retirement of D. H. Hey and W. A. Waters, and *Proceedings of the XIII international congress of pure and applied chemistry*, vol. 4 (London: Butterworths, 1971).

Special topics

Articles and books specially relevant to specific sections of this text (shown in brackets) are as follows.

A. R. Forrester, J. M. Hay and R. H. Thomson, *Organic chemistry of stable free radicals*, London: Academic, 1968. (Ch 3).

J. M. Tedder, *Q. Rev. chem. Soc.*, 1960, **14**, 336. (Abstraction reactions, Ch 5).

R. S. Davidson, *Q. Rev. chem. Soc.*, 1967, **21**, 249. (Abstraction reactions, Ch 5).

E. G. E. Hawkins, *Organic peroxides*. London: Spon, 1961. (Ch 4).

G. Scott, *Atmospheric oxidation and antioxidants*. Amsterdam: Elsevier, 1965. (Ch 5).

J. I. G. Cadogan and M. J. Perkins in *The chemistry of the alkenes*, (S. Patai ed). New York: Wiley, 1964. (Ch 6 and 9).

J. C. Bevington, *Radical polymerisation*. London: Academic, 1961.

G. H. Williams, *Homolytic aromatic substitution*. Oxford: Pergamon, 1960.

G. Sosnovsky, *Radicals in synthesis*. New York: Wiley, 1968. (Ch 8).

T. L. Gilchrist and C. W. Rees, *Carbenes, nitrenes and arynes*. London: Nelson, 1969. (Ch 9).

J. C. P. Schwarz (ed), *Physical methods in organic chemistry*. Edinburgh: Oliver & Boyd, 1965. (Ch 10).

A. Carrington, *Q. Rev. chem. Soc.*, 1963, **17**, 67 (esr Ch 10).

D. J. E. Ingram, *Free radicals as studied by electron spin resonance*. London: Butterworths, 1958. (Ch 10).

CIDNP: H. R. Ward, *Accts chem. Res.*, 1972, **5**, 18; R. G. Lawler, *ibid.*, p 25 (Ch 10).

K. U. Ingold and B. P. Roberts, *Free radical substitution reactions*. New York: Wiley, 1971.

M. Bersohn and J. C. Baird, *Introduction to electron paramagnetic resonance*. New York: Benjamin, 1966. (Ch 10).

Index